Jerome Cardano
(1501–1576)

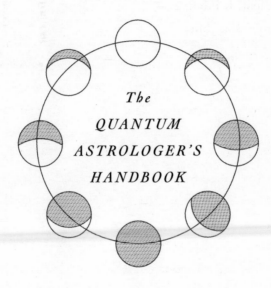

The

QUANTUM

ASTROLOGER'S

HANDBOOK

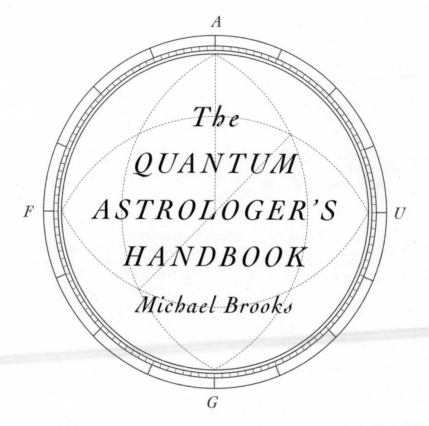

A

F   *The*
QUANTUM
ASTROLOGER'S
HANDBOOK   U

Michael Brooks

G

SCRIBE

Melbourne • London

Scribe Publications
18–20 Edward St, Brunswick, Victoria 3056, Australia
2 John St, Clerkenwell, London, WC1N 2ES, United Kingdom
3754 Pleasant Ave, Suite 100, Minneapolis, Minnesota 55409 USA

First published by Scribe in Australia and UK 2017
Reprinted 2017
Published by Scribe in North America 2019

Typeset in Adobe Garamond 11.65/17pt by the publishers
Printed and bound in the UK by CPI Group (UK) Ltd, Croydon CR0 4YY

Scribe Publications is committed to the sustainable use of natural resources and the use of
paper products made responsibly from those resources.

9781947534810 (US hardback)
9781925322408 (Australian paperback)
9781911617358 (UK paperback)
9781925548280 (e-book)

Catalogue records for this book are available from the National Library of Australia
and the British Library.

scribepublications.com
scribepublications.co.uk

To Phillippa, my Lucia

I was ever hot tempered, single minded, and given to women ... cunning, crafty, sarcastic, diligent, impertinent, sad and treacherous, miserable, hateful, lascivious, obscene, lying, obsequious ...
JEROME CARDANO

Cardano was a great man with all his faults; without them he would have been incomparable.
GOTTFRIED LEIBNIZ

Prologue

It is 6 October 1570. In England, Guy Fawkes is a newborn baby in his mother's arms and Queen Elizabeth I is feeling the sting of her excommunication from the Catholic Church. In Italy, the once-great Jerome Cardano, now sixty-nine years old and feeling it, is also about to fall foul of the religious establishment.

He is in Bologna for a meeting of the city's syndics, the governing officials who pronounce on civil law. Jerome hopes to persuade them of his innocence — that he has not, as the Milanese College of Physicians has suggested, committed sodomy and incest. Forbidden to enter Milan to plead his case, his only hope is the Bolognese syndics. But his hope is misplaced and he seems to have no concept of just how impossible his position has become. In the public's eyes, he is now a madman. In Milan, he was spotted begging for alms at the gates of the College of Physicians — where he once held the office of rector. There are moments when Jerome is overcome by his new misfortune, by his hunger, and by the ignominy of his position, and is found loudly cursing in the streets. It doesn't help that he has taken to wearing a gift given to him many years earlier by the Archbishop of Scotland: a belted plaid that he wraps around his waist and secures with a leather belt, throwing the rest of the heavy mottled cloth over his shoulder. No one in Italy has ever seen — let alone worn — anything quite like it. Who can blame the locals for laughing?

How the mighty fall. Just two decades earlier, this man was summoned all the way to Edinburgh to treat the Archbishop's asthma. On his long journey to Scotland, the physicians to the French king had organised for Jerome to speak at a series of conferences in Paris. Then, while in Edinburgh, the courtiers of the young King Edward VI of England had begged him to come to London and provide a medical consultation for the ailing royal youth. Not satisfied merely to tap Jerome's medical skill, they prevailed upon him to construct the royal astrological chart. He left Edinburgh a rich and celebrated man; he left London even richer. On his journey home, he travelled via all the major cities of Europe, entertained by noblemen and the ambassadors of the Holy Roman Emperor.

Now he has no money to pay for lodgings and spends his nights in an abandoned hovel where the wind whistles through the gaps in the walls. What is left of the roof creaks ominously above his head. Every evening, before he goes to sleep, the celebrated physician, the royal astrologer, the inventor of numerous machines and mathematical abstractions — among them probability theory — eyes the rotten beams. He attempts a calculation of the likelihood of the building's collapse. There is a part of him that would welcome such a swift end.

But morning comes and the building still stands. His stomach empty and groaning, Jerome emerges warily into the light and looks down the street. He has woken in a good mood. There is a lightness in his stride as he sidesteps a mangy, sleeping dog — he has developed a phobia of dogs that he will attempt to explain in the pages of his autobiography — and turns towards the city centre. Today, he will see the Bolognese syndics and they will listen to him. These are not like the petty, sour-faced goats that rule over Milan. From tomorrow, he will be permitted to earn his living again. And then, across the road, he sees someone staring at him. At first, the disfigured, bearded features of Nicolo Tartaglia are not clear. But then the man known as The Stammerer steps forward, and with him steps a cohort of the city guard, their armour gleaming in the early morning sun.

'There he is,' Tartaglia says. His words are barely discernible, so profound are the childhood injuries to his face. But the gleam in his eye is unmistakable. 'Arrest him.'

As the guard moves to cross the street, another figure is revealed. Watching with a cold, intense gaze is Aldo, Jerome's youngest son. Slowly, the young man turns and walks away. But not before the father sees a sly smile, a grin that celebrates a long-sought revenge, bloom on the face of his only surviving child.

$$\psi$$

Have you ever wanted to understand the universe? Once that desire burns away at your soul — really burns — there's no going back. That's why some people dedicate their lives to physics. Or to philosophy. Or to Buddhism. Or to mathematics. They are all searching for answers. I am not saying that they are all ultimately following the same path — I know which I think is the best bet — but none is able to satisfy everybody.

I chose physics as my path to enlightenment. Some prefer the teachings of Jesus. Others go for Krishna or Kabbalah. My friend Jerome Cardano — indulge me, for we have spent a lot of time together — opted for astrology. He didn't ever really trust it, though. Jerome used to worry at astrology, to work it hard, to ask difficult questions of it. I'm not convinced everyone does the same, even with physics — a predicament that provides much of the reason for this book.

I am a physicist. My expertise, such as it is, is in quantum mechanics, the theory that describes how the world works on microscopic scales. My interest in Jerome arises from the fact that he used his sharp mind to unearth the mathematical pillars on which quantum theory, our most successful scientific guide to the universe, is founded. Astrology and quantum physics rattling around in one Renaissance skull — who'd have thought?

Jerome would be happy that I am introducing him to you: his work, his mind, and his life. He always wanted to be famous; by the age of twelve, he had decided to dedicate himself to creating something that would bring him lasting renown. That you know next to nothing about him points to one of his many dashed hopes.

He hoped also to make his fortune at the gambling table. Despite inventing probability theory for just that purpose, he gambled away his marital bed and all his wife's jewellery. Then there was his hope that his wife, Lucia, would live a long and happy life. For all the good doctor's successes in treating others, he could do nothing to halt her death after just fourteen years of marriage. He hoped that his elder son would be a successful physician. Unfortunately, Giovanni's marriage into a family of gangsters made that particular aspiration particularly optimistic, and the young man's execution for murder, a plot twist that broke Jerome's heart, put an end to that hope. Jerome also hoped for grandchildren, but ended up raising only the grandchild of a man who tried to ruin him.

The one thing for which he held no hope was probably his most important and lasting creation. It is the square root of a negative number, something we now call the imaginary number. Though it initially seemed like nothing more than a strange mathematical abstraction, it has turned out to be essential to understanding how the universe holds together.

It was a privilege to be the one to tell him.

You're probably thinking that I have lost my mind. You might be right. My obsession with Jerome has, in the last few years, taken me over. I have a mind that has been schooled in quantum physics and trained to think rationally, dissecting facts and ideas dispassionately. And here I am, not only celebrating a Renaissance astrologer, but talking as if we are contemporaries.

To me, though, it makes sense. I talk to Jerome. He talks to me. These conversations take place in my head, true, but they are informed by his writings, and by things written about him. We are intellectual contemporaries. We are both rational, both seeking to understand the universe, both convinced that nobody has a good grasp of it yet. We both believe that space and time —

time, in particular — are not what you and I have been raised to believe they are. So, yes, this book is not quite what you have come to expect from a science writer with my training and history. But I can't help that. In my head I have visited Jerome in his prison cell. And maybe it is not just in my head. Within the books that Jerome wrote after his release I see unmistakable traces of my visits.

Perhaps you should walk away before I drag you into this madness.

Chapter 1

Jerome has been in this cell for eleven weeks since his arrest. The weather has turned cold and damp, and he struggles to keep himself warm. Until yesterday, he had no idea why he was in prison; no one would tell him anything. The hunched man who brings fresh straw every day refuses even to look at him. The tall, thin boy who brings the food smiles as he sets the bowl down on the writing desk, but has nothing to offer but a shrug in response to Jerome's questions. Yesterday, though, a new face entered the cell. When the guard turned the key and opened up the door, the stranger took one step in, threw down the yellow robe, smirked, turned, and walked away. And then Jerome knew.

He puts down his pen and turns his attention to the robe that now smothers his emaciated frame, pulling at it as if it burns his skin. It is embroidered with demons who are using forks and fiery flames to torture hapless, grimacing men. Jerome knows its significance: it is what the heretics wear when led to the stake.

We are in December now. Christmas is coming and the last light of the day is fading ever earlier. The cell is small and dark, with a window big enough only for a small boy to climb through.

It has been many decades since Jerome was a small boy. He is sat behind a rickety desk — a small mercy afforded to him by the authorities. Perhaps they hope he will write more blasphemies and make their case easier to prosecute.

Jerome looks up from the robe and stares at me through the gloom. He is not sure if I am an apparition. I am not sure, either. Eventually, without shifting his gaze from my eyes, he puts his fingers to his mouth and pulls out a small emerald. It is held on a chain around his neck. He lets the stone fall to his chest.

'Do I know you?' he says. His voice is thin and high pitched — reedy — entirely unbecoming to a man in his seventh decade.

'I don't think so,' I say.

'Did they send you to spy on me?'

'They?'

'My Inquisitors.' He pulls at the yellow robe again.

'No.' I break from his gaze and survey the cell again. 'I'm just here. As far as I can tell.' It is all I have to offer.

It seems to satisfy. 'Oh,' he says. He picks up the pen again. 'Well, welcome, then.'

'Are you writing to Archbishop Hamilton?' I say.

He stares at me. 'Why would I do that?'

'To ask for his help.'

Jerome shakes his head. 'He will be dead by now,' he says. 'I'm sure of it.'

I have read the history books and I know the truth. 'He's not. Your treatment was more successful than even you might have hoped.' I hesitate, unsure whether this constitutes some breach of the rules of engagement. I decide I don't care. Nobody has told me the rules. 'You should write to him,' I say.

And that is how, I like to think, I came to save the life of Jerome Cardano.

ψ

It is going to be difficult to convince anyone that I saved Jerome. Jerome was born in 1501 and I in 1970. Bound as we all seem to be by time's arrow, I can see that there are problems with the concept. I may already be coming across as an unreliable narrator. But, before you judge, first learn something about where I'm coming from and — more importantly — familiarise yourself at least a little with the ideas of quantum theory.

According to our best description of the atomic and subatomic world, atoms and their constituent particles are able to exist simultaneously in two places at once. It's there in the theory and we've seen it in experiments. They can even exist at two different moments simultaneously. So, even as they gather to form my body, their notion of time and space is utterly different to the one I experience. And so, I ask, why shouldn't I be in two places and epochs at once?

I'm playing with you, of course. I *am* an unreliable narrator. That's the whole premise here. But aren't we all? After all, I've already mentioned my 'experience' of time as if I know what that means. All I can tell you about that particular phenomenon is that my experience involves my consciousness — something that scientists can't even define, let alone explain. If quantum physics is slippery, it's nothing compared to the minimal friction you'll encounter when you try to pin down a neuroscientist on consciousness.

One of the problems is that consciousness is entirely subjective. I believe I am conscious; I have no way of telling whether you are. You, to me, are therefore an unreliable narrator. A narrator is only reliable when we can corroborate their version of events. We take the view that if several people agree on a narrative arc, it is probably a trustworthy description of how things happened. But how can I trust anyone else if I don't know what's going on — if anything — in their head? What's more, it doesn't mean that other things — things no one mentioned — *didn't* happen. Even the agreed narrative may not tell the whole truth.

I certainly cannot corroborate Jerome's version of events. I can only go by what he says — and what he says is sometimes odd. I first came across him some years ago now, when researching a book about how science works. I was writing a chapter about the origins of scientific creativity and was seeking out examples of strange sources of inspiration: hallucinogenic and dream states, daydreams or poetry-inspired visions, and so on. Most scientists chose to hide these questionable sources. But not Jerome.

He invented the mechanical gimbal that was to make the printing press possible. His idea led to the 'Cardan joint' that takes the rotary power in the driveshaft of your car's engine and allows it to be transmitted to the front and rear axles. We have already mentioned the mind-bending imaginary numbers that are multiples of the square root of -1 and the original mathematics of probability. He pioneered the experimental method of research in areas as diverse as medical cures for deafness and hernia, cryptography, and speaking with the dead (forgive him, these are not strictly scientific times). Jerome's autobiography details some of these achievements, yet when he documents their source he says they came from 'the ministrations of my attendant spirit'.

Here we would say he is indulging in unreliable narration. We tend not to believe in visiting spirits, especially those that impart scientific insights. So this is surely a lie, or the raving of a disturbed mind? Jerome's father also had a spirit visitor, as it happens. As a trained scientist, I should probably put all this down to a genetic predisposition to psychosis or schizophrenic delusions. Yet despite, or perhaps because of, this I became quietly fascinated. I read everything on Jerome that I could find. The vast majority of his four million words of writings (four million!) are only available in Latin (not my strong suit), but there are a few biographies in English. A couple were written in the nineteenth century. A Norwegian mathematician called Øystein Ore published another in 1953, focused on Jerome's probability work. There's a more general biography from 1969, written by a jobbing journalist called Alan Wykes. More recently, some academic scholars have dissected Jerome's astrological studies and his medical work. All this seeped into me, permeating my thinking and my imagination, and mixing itself with my experience and my knowledge. Then it set hard in my brain as thoughts and imaginings about the possible, the probable, and the unlikely. It became a new narrative for me, as compelling as quantum theory and just as unreliable. Jerome and I are now inextricably entangled across space and time, just like the photons that spooked Einstein so badly.

Photons, I should explain, are the fundamental particles of

light and other radiation. They travel at — unsurprisingly — the speed of light, which is the maximum speed of anything in the universe. In his special theory of relativity, Einstein showed that travelling at the speed of light is equivalent to halting time. That means photons do not experience time, as such. However, that didn't stop him balking at entanglement, arguably still quantum theory's most shocking revelation.

This is the discovery that you can cause two photons (or any other quantum particles) to interact so that their properties become shared between them. You can then separate the pair by half a universe, do something to one and immediately see the effects of your action in the properties of the other. Einstein dismissed entanglement as proof that quantum theory must be somehow incomplete, deriding it as 'spooky action at a distance'. We know now that entanglement works across time as well as space. We'll get into this later. All I'm saying is, given that Jerome is spooking me now, perhaps I spooked him then.

For the moment, let's go back and visit the beginning.

$$\psi$$

It is astonishing — a testimony to his tenacity — that Jerome is born at all. This is the beginning of the sixteenth century and Italy is in its Renaissance period. As far as most of its inhabitants are concerned, this is not the Renaissance Italy you are thinking of, with its glorious legacy of art and culture. True, Leonardo's creativity is in full flow, and in a few months Michelangelo will return to Florence and begin work on his statue of David. But this Italy is a patchwork of regional states, broken by centuries of petty internal conflict and civil war, and rotten with plague, poverty, and superstition. Its rulers are in thrall to a succession of bloated, self-aggrandising popes. Living through the harsh reality of everyday life in Italy's Renaissance is not glamorous, and that is why, when Chiara Micheria, a fat little widow with a short temper, realises she is pregnant again, she seeks out an apothecary.

Having gained herself a consultation, she asks the apothecary how best to induce an abortion. The church's newly published witch-hunting manual, *Malleus Maleficarum*, has pronounced abortions the work of the devil, and midwives inducing them are to be treated as witches. But, however dangerous the move might be, Chiara feels herself in no position to deal with yet another bastard child. After all, she already has three brats of dubious parentage.

Truth be told, Chiara is a woman of somewhat loose morals and everyone around knows her brood may not have been her husband's progeny. Not that he is around to protest. A few years ago he was murdered in a bar brawl, the result of being caught cheating at the card table. To Chiara's credit, the father of this latest baby *is* known, at least — and almost respectable, too. Fazio Cardano is a mathematician and jurisconsult. He is not much to look at, though: fifty-six years old and toothless, with shoulders that are stooped and rounded from hours hunched over Euclid's books on geometry. And even if love were blind, marriage is not an enticing prospect for these two. The couple argue constantly and neither wants to bind themselves to the other just because of a child. What's more, Milan is beset by new cases of the plague. Chiara is a practical woman, which is why she has left Fazio to take his chances, gathered up her children, and headed twenty miles south to plague-free Pavia.

It is here that the apothecary recommends she drink a poison to bring on a miscarriage. She follows his advice to the letter, but it doesn't work. Neither do the next two doses. Despite all her efforts and expenditure, on 24 September 1501, after a three day labour, a nursemaid violently pulls a child with a head of curly black hair from between Chiara's thighs. At first, it seems the reluctant mother is in luck: the child appears to be stillborn. However, the nurse puts him in a bath of warm wine and Jerome Cardano kicks into life.

This is not the last time that Jerome stares death in the face. Less than two years later, while he is still being suckled by a wet nurse, an outbreak of plague kills his half-brothers and half-sister. The nurse, too, catches the plague and dies. Her infected milk brings him out in five carbuncles, including one on the end of his nose, which scar and mark him for the rest of his life. But he survives, growing into a sickly boy who is prone to long bouts of illness and is often confined to his room in the family home in Milan's Via del Rovelli.

The illnesses are frequently severe, but they never defeat him. Perhaps he is made hardier by the rough treatment he receives at the hands of his parents, both of whom beat and whip him. From the age of five, Jerome accompanies his father on visits to clients, acting as his book-carrying page. Occasionally, when Fazio needs to consult a volume en route, he orders Jerome to stand still and places the book on his crown. If Jerome moves, disturbing his father's concentration, he receives a blow to the head. It is dark testimony to the treatment he gets from his mother that Jerome recalls his father as the 'more loving' of the two.

It's not all hardship though. Such journeys give Jerome experience of the best his world has to offer. He sits in on discussions between his father and Leonardo da Vinci. He sees Leonardo's *The Last Supper* on the refectory wall at the convent of Santa Maria delle Grazie, long before it decays. The artist experimented with paints that lasted only a couple of decades before they began to flake and fade. Seeing it later in life, Jerome called it 'blurred and colourless' in comparison with the glorious painting he had seen as a boy. I daren't tell him about Napoleon's troops peppering it with stones and horse dung.

For all the occasional glamour, serving his father fixes Jerome's gaze on a different path. Fazio is grooming his son for a legal career, but in vain: by the time he is eight years old, the boy has decided that legal books are too heavy and legal arguments too dull and inconclusive. Numbers, geometry, and medicine are much more attractive. By the age of twelve, he has enough understanding in arithmetic to read the works of the Arabic alchemist Jabir ibn Hayyan and has developed a theorem that allows him to calculate the distance between stars from their latitude and longitude. The theorem is almost certainly riddled with errors. Nonetheless, the young Jerome's desire to explore the cosmos is making itself clear, as is his lust for lasting fame. 'I desire to defend myself from obscurity,' the precocious twelve-year-old writes in a letter to a friend.

$$\psi$$

'Five hundred years later, people still know your name,' I tell him. He looks unconvinced.

He steeples his hands together, laying his elbows on the splintered wood of the desk. The apex of the steeple covers the pox scar on his nose — the pose looks practised, useful for hiding this small disfigurement.

'It's what you wanted, isn't it?'

He gives me no response. Just a continued fixed stare. He appears to be sucking on the emerald again. It is hard to know how to take this forward. I want to flatter him, to convey my admiration. I don't want to tell him that he is actually overwhelmingly unknown. Really, it is only a few mathematicians and a smattering of historians who have any familiarity with his work. Every day, hundreds of millions of people drive around in cars whose power transmission depends upon a Cardan joint, but they don't know anything about the man himself. Also, I'm wary of opening up a discussion of cars and internal combustion engines. That might be a lot to take in for a man who has just learned he is to face the Inquisition.

'What are they accusing you of?' I ask, nodding to the door that leads out of the cell. I ask as if I don't know.

Jerome's eyebrows lower and his eyes narrow. He stares at me in open suspicion. Finally, he takes the stone from his mouth and breaks his silence.

'My father told me he was visited by a familiar spirit the whole of his life,' he says. 'Are you the same one?'

'Is that what you think I am?' The idea makes me laugh out loud. 'A spirit? Your guardian angel?'

He stares at me, confused, then begins to smile, an endearing but ugly sight that lifts the pox scars on his face out from behind the hairs of his beard. 'It was *you*,' he beams. '*You* sent me the warnings. About my mother's death, and Giovanni's marriage, and ...'

He stops because I am still laughing. This is the trouble with us human beings. We have no ability to confine ourselves to the realities of the natural universe. As soon as anything remotely inexplicable happens, we reach for the supernatural.

ψ

Perhaps this is a good time to tackle the astrology question. Here's the simple truth: I could not be such good friends with Jerome if I did not consider him a rational man. He is not one to attribute supernatural causes to everything that he can't explain. In fact, he celebrates the strange and wonderful abilities of the human mind, its seemingly endless capacity to reason, imagine, and invent. He records his encounters with a variety of technologies and innovations that have impressed him over the years, things such as a hinged, hooked hand that would close on a thief's fingers if he tried to take money from a purse. 'Our age is prolific of distinguished and very great inventions,' he says in *De Subtilitate* (*On Subtlety*), a book he published in 1550, aged forty-nine.

Nothing impressed him more than the skills of a great conjuror. He marvels at the card tricks that confuse, then amaze, the onlooker. He doesn't understand how the conjuror can cause his audience to choose specific cards from the pack. 'It was too extraordinary for us to follow by human deliberation … unless he had sometimes asked us to take out a number of different cards, I would have suspected that he had supplied a pack consisting of cards of the same kind,' he says. For all the mystical thinking of his epoch and his own belief in the reality of the supernatural, he sees no dark influence at work. 'The evidence was that it all was the work of a conjurer rather than demons,' he writes.

If the card sharps were entertaining, the Renaissance showmen were astonishing:

> *There is no end to the inventions of this skill — shifting things, hiding them, swallowing them, emitting floods of fluid from one's eyes or forehead, producing nails and string from one's mouth, chewing glass, piercing one's arms and hands with a spike, uniting iron chains while the links remain intact — indeed, a greater feat, I have seen three rings thrown up and coming down interlinked, though they were unbroken and separate before and while being thrown up … They show a child without a head, and the head without the child — but they are all alive, and the child comes to no harm in the meantime.*

Then there are the rope walkers, who would climb steep lines from the ground up into a tower, often while balancing something — or someone — on their shoulders. Jerome assures us in *On Subtlety* that there is no occult power at work. It is, he says, simply a demonstration of the extraordinary feats to be achieved when we gain an understanding of how nature is. 'It is in fact magic, when something extraordinary is done on natural principles, concealed though these are,' he says.

I am telling you all this because I want you to understand he is no different to you and me, even if he does believe in the influence of the stars and planets over the lives of humans. Such a belief is a completely tolerable idea at his stage in history and it is still surprisingly inviting to an intelligent human mind. I know this because I tried it out.

I decided quite early on that, if I am to understand Jerome, to know him a little, I should dabble in astrology. To this end, I had a couple of readings done. One was by an Indian astrologer called Vishal. He knew only my name, and date and place of birth, and didn't tell me anything terribly profound until I asked him about the car I had just bought. His astrological charts were all on his laptop computer (a pleasing conjunction of ancient and modern) and it didn't take him long to tap something car-related into the program.

'I see two cars in your future,' he said.

I laughed. 'Does that mean I've bought a dud?'

'I can't tell you that,' he said.

Apparently, Vishal just saw two cars. And two weeks later, my mechanic was working on the one I had bought when he found a significant problem. He informed me I had indeed bought a dud. I wish I could tell you it was an issue with the Cardan joint, but it was a simple case of hidden but chronic chassis corrosion. He advised me to return the car, get my money back, and buy another — preferably from a reputable dealer.

My other reading was by a woman called Sue. Perhaps I had

been primed by Vishal, but some of her pronouncements seemed uncanny. Then, however, I had a psychologist listen to a recording of the session and he was gloriously unimpressed. Sue was performing the classic generalisations designed to connect with almost anyone of my age and gender, he said. Vishal, he added, had just been lucky with a stab in the dark about the car.

I felt suitably chastised. But it made me realise how easy it is to believe that hidden causes lie behind everyday events.

Jerome's investigations of astrology were, to him, entirely rational. In a worldview formed by the biblical scriptures, heavenly portents are a given. The book of Revelation declares that the heavens are like a scroll of parchment — so why not see what's written on it? Genesis tells us that the stars and the moon were given 'for signs and for seasons'. There are moments in Israelite history when we are told the sun stood still in the sky. The prophets saw all kinds of celestial signs associated with divine judgement: Isaiah tells us the sun will be 'dark at its rising'; Joel tells us 'the stars withdraw their shining'. True, there are passages condemning astrology used for divination, but that stricture is about attempting to know the mind of God. The writers clearly want us to be aware of God's control over nature and its betrayal of his mood in signs written on the sky.

Even within this paradigm of reasonable astrology, Jerome is something of a progressive, imbued with a rising scepticism. Take that as a mark of his fine mind: the scepticism arose despite his conservative upbringing. Jerome's father taught him the principles of casting horoscopes and he was often in the company of men — always men — discussing the art of interpreting the celestial sphere. Which makes it all the more remarkable that the young Jerome writes this to a friend in 1519, at the age of eighteen:

> *It was very well for my father to wring his hands at the*
> *issue of fortune determined by the heavenly bodies on the*
> *houses where we set our homes; but I recall thinking that*
> *the distance was very great for such an issue to be made, for*
> *might not the sun shine on Cathay when rain fell upon*
> *Padua? And if the biggest star of all could not stretch its*
> *warmth thus far, how could the smaller ones seem to malign*
> *from such a height? Now I know I was in error; but it was a*
> *childish curiosity and not without the gleanings of a cynic.*

Centuries later, we science lovers consider ourselves to have a more sophisticated perspective. There is no physical force — and certainly not one associated with the stars or planets — that might influence our personalities or life events. Yes, there are forces that act at a distance — gravity and electromagnetism, for instance — but they are weak and ineffectual, even over the kinds of distances that stretch between Earth's countries and continents. What force could possibly exert influence on us from the planets or stars?

But even in Jerome's time, it was not a modern thing to question astrology. Cicero and Augustine had offered criticisms. Logical thinkers had pointed out that the time and date of birth could not determine character and destiny, as astrologers claimed, because twins often had very different characters and destinies. The Genesis account of Jacob and Esau makes that abundantly clear. What's more, astrologers were adept at self-examination.

Ptolemy, the Greek-Egyptian mathematician, astronomer, and astrologer, was a vicious critic of his own predictive powers and their limitations. It was acknowledged that, in a world created only a few thousand years before, as the contemporary cosmology had it, there hadn't been time for some of the conjunctions and other phenomena to occur repeatedly. Thinkers acknowledge that if the celestial events had happened only once in recorded history, any claims about their significance relied on entirely spurious single-point observations.

Jerome remained conflicted. The stars and planets are not *too* distant to exert some influence, he suggested. It is clear from his other writings that he is convinced an influence does exist. However, he admits he has only the vaguest idea of what is going on and that it may not be exerted over everything in life. 'A man is a fool who attaches too much meaning to insignificant events,' he says in his autobiography, *The Book of My Life*. He was seventy-five years old then and clearly still believed the heavenly portents were signs of something — they were just incredibly difficult to read. 'There were stars which threatened, from every aspect, my death, which all declared would be before my forty-fifth year,' he writes, 'all vain findings, for I live, and I am in my seventy-fifth year! It is not the fallibility of the art; it is the inexperience of the artificer.'

ψ

Jerome's views on astrology mirror our own on quantum physics. In quantum experiments we see things appear in two different places at once, or exert an instantaneous influence over something that is half a world away. We cannot make sense of it, but we don't dismiss it as ridiculous. We have the evidence of our experiments, after all, just as the astrologers have the 'evidence' of experience.

For three years, I worked in a research laboratory studying quantum physics. At the end of those three years, I produced a report of my research: my PhD thesis. It sits on the shelf above my desk, bound in blue with the title embossed in gold: 'Quantum mechanical behaviour of superconducting weak link capacitor circuits in the range from 9K to 0.3K.' It describes a project to take a ring of niobium metal and make the electrical current circulating within it travel in two directions at once. Let me be clear: this is not two different currents, but one, doing two things simultaneously.

The phenomenon is known as superposition. I was not the first to create a superposition in a niobium ring — far from it (though I was the first to do it at the particularly low temperatures to which my niobium ring was cooled). Nor was I the last; others have followed up the work. But we still don't really understand it.

The explanation of superposition — such as it is — is more commonly laid out in terms of the 'double slit experiment'. It demonstrates that matter can simultaneously exist as discrete objects, like an archer's arrow, and as waves, such as those that travel through a large, continuous body of water.

Imagine an archer trying to fire arrows into the watchtower of a fortified medieval castle. His only hope of getting his arrows to the inside is by firing at one of the two narrow arrow slits set two metres apart in the wall. Assuming he is good at his job, his arrows will hit the wall behind the arrow slits. There will be a cluster of arrows at two distinct points on that wall.

Now imagine a trebuchet that throws a huge bucketful of water at the wall. The water will go through both slits and the two portions that emerge into the room will hit each other as they travel towards the far wall. That means they affect each others' paths, so there won't be two distinct splashes on the wall, but a messy wet patch that stretches across its extent. It might be a bit wetter at the points directly behind the arrow slits, but much of the rest of the wall will be wet, too.

We can imagine this because we know how single objects or particles behave and we know how things like large bodies of liquid behave. What we also know now is that there are things that behave like both — and neither.

Physicists generally describe the double slit experiment in terms of light. The light is fired at two very narrow slits in an opaque barrier. The far wall on the other side of the barrier is a light-detecting screen.

When English scientist Thomas Young shone light on a pair of slits in 1803, he found that the light hit the screen giving a peculiar pattern of bright and dark areas. He explained this in terms of two light beams emerging from the slits (one from each slit) and 'interfering' with each other. Interference was a known property of waves. When two water waves, for example, travel in the same direction, they will spread out and overlap. If the crest of one wave coincides with the crest of the other, their strength is doubled, and a larger crest is created. Similarly, if the wave troughs coincide, the trough is deepened. Where a crest meets a trough, the wave is destroyed and the water is, effectively, flat.

The pattern of light and dark 'fringes', to Young, was long-sought proof that light is a wave, not a particle. An old debate was finally settled.

Except that it wasn't. Early in the nineteenth century, Einstein showed that light exists as particles we now call photons. He won a Nobel Prize for his efforts and a few decades later we learned to control lasers that could emit one photon at a time. That changed the face of physics forever.

Repeat Young's experiment with just a single photon fired at the slits and something utterly extraordinary happens. You have to replace the light-detecting screen with one that permanently records the impact site of each photon by changing colour, say from black to white, but it is worth the effort. You also have to be patient: with just one photon hitting the screen at a time, it takes a little while for the miracle to emerge. But emerge it does.

Eventually, the screen has turned from black to white in exactly the same places as it would if two beams of light were emerging from the slits. There is still interference, despite the fact that there is only one photon in the apparatus at any one time. The only explanation is that the single photon has somehow travelled through both slits simultaneously and then interfered with itself. This experiment — which has been conducted countless times without any failures — contains what the celebrated physicist Richard Feynman called 'the only mystery' of quantum physics. The mystery is that, given a choice of two paths, a quantum particle takes both. It happens with single electrons (Claus Jönsson of the University of Tübingen was the first to observe that, in 1961) and with larger particles, such as atoms and molecules. It happens with the large body of quantum material (known as a Bose–Einstein condensate) that makes up the electrical current within my ring of niobium. Yet none of these occurrences make sense.

What makes it even more confounding is the fact that, if you try to see which path the particle took by placing some kind of detector on one of the slits, the interference pattern goes away and you go back to two distinct clumps. Under observation, continuous waves turn back into discrete particles. It is as if watching the water emerge from the arrow slit makes the water change its fundamental character and suddenly behave like the archer's arrows. In the quantum world, it seems, the weirdness doesn't like to be watched.

This is entirely within the theory's predictions. Quantum theory is our most successful mathematical framework and its predictions have never been wrong. However, it is utterly bankrupt when it comes to explaining the physical world. It tells us what we will see in any given quantum experiment, but leaves the 'interpretation' — a description of what is *actually* going on — entirely up to us.

So, what happens between the time when the photon enters the apparatus and is detected at the other side? When does the photon begin its double existence? If those slits could talk, what would they tell us? We don't yet know.

The interpretations of quantum physics are the tales we have constructed around this simple experiment, and any one of them might be true. One interpretation — explanation, if you like — says the photon simply isn't real until you detect it at its final resting place. That means you can't actually say that it went through both slits; in the absence of detection, it didn't have any existence in the experiment. Another interpretation says there is an undetectable 'guiding wave' that determines the photon's path through one slit, while leaving a false trail that suggests it went through both. Another is that there is a branching of worlds where, in one world, the photon went through one slit and, simultaneously, in another world, it went through the other. The interference pattern we see is the 'crosstalk' between these worlds.

What do you make of this? Are you comfortable with things pinging into existence only when they hit a detector? Can you stretch to a false trail created by an undetectable wave? And what about these parallel worlds? Does that seem like it could ever be the simplest, most plausible explanation for the outcome of an experiment?

Of course not. So we delve deeper, convinced that the fault lies with our poor grasp of reality. 'Quantum theory does not trouble me at all. It is just the way the world works,' the great physicist John Archibald Wheeler once wrote. 'What eats me, gets me, drives me, pushes me, is to understand how it got that way ... the quantum is the crack in the armor that covers the secret of existence.'

$$\psi$$

For the young Jerome, curious and sharp-minded, the only way to learn the secret of existence was to practise the art of astrology. This he did throughout his life, quickly becoming expert at getting under the skin of his subject — and gaining the attention of paying customers.

The speed of his advances is illustrated in the difference between his first and second astrological publications. The first, published at his own expense when he was thirty-three, was a 'Prognostico'. Typical of the genre, it comprised a series of short- and long-term predictions about developments in anything from religion to politics to the weather. Such pamphlets were commonplace — Europe was crawling with astrologers. No amount of official derision would dampen public enthusiasm for foretellings. The Archbishop of Canterbury dismissed Michel de Nostredamus's 1558 prognostication as a 'fantastical hotch-potch', for example, but a vast swathe of Europeans, from merchants in the market to princes in their palaces, acted on Nostradamus's words.

Jerome was keen to divert some of this attention (and money) his way. To distinguish himself from other, lesser practitioners who were 'defiling this noble science', as Jerome put it, he announced himself as relatively unknown. Not, Jerome added, because he was a poor practitioner of the art, but because he was not one of those people who is happy to achieve fame by saying what noblemen want to hear.

Having set out his stall with this humblebrag, he blinds the reader of his Prognostico with details of offset axes in the celestial spheres and the precession of equinoxes. There is no shortage of technical detail by which he establishes his scientific credentials. He stops short of predictions about wars because 'there is no part of astrology harder than this one, and yet the bulk of these crazy diviners speak more boldly about it, in their bestiality, than about anything else.'

Jerome later refers to his works of this time as 'astronomical tables', but however scientific Jerome wanted his astrological works to appear, they were primarily published to make money and create a reputation. As a backup, he was careful to record himself as 'a doctor from Milan' on the title page — an advertisement for his other line of business. He was, at this stage, not yet licensed to practise medicine by the Milanese College of Physicians and, without a licence, he couldn't officially see patients within the city. Yet he needed to scrape together a living and was open to offers. Jerome knew he had his work cut out making it as an astrologer in such a controversial and overcrowded space. That's why he followed a different path with his second publication: *Supplement to the Almanach*. Published four years later, in 1538, its unique selling point is its inclusion of a primer on astronomy. 'One who wishes to attain knowledge of the stars must begin with knowledge of the planets,' Jerome explains. And so he lays out the movements of the known planets and instructs the reader on how to find each of them in the sky. He explains how to find the pole and all the constellations of the Zodiac. He gives tips on remembering all the various positions. It is an unprecedented popularisation of astrology, removing the barrier between amateur and professional; Jerome is extending an open invitation to anyone wanting to understand the secrets of the stars. He is, according to the twenty-first-century scholar Anthony Grafton, 'a sixteenth-century counterpart to Patrick Moore or Carl Sagan'.

All he needs from his client is humility and tolerance. 'This alone I ask you, O reader, that when you peruse the account of these marvels that you do not set up for yourself as a standard human intellectual pride but rather the great size and vastness of earth and sky.' It is Jerome's way of excusing his approximations and errors: astronomer and astrologer cannot hope for absolute precision when dealing with the near infinite phenomena of the celestial sphere. That said, Jerome considers himself well within the limits of the plausible. 'And, comparing with that Infinity these slender shadows in which miserably and anxiously we are enveloped, you will easily know that I have related nothing which is beyond belief,' he says.

A second volume of this publication — *Corrections to Errors of Time and Motion* — explains the latest thinking about the motions of the planets, offering corrections to errors in accepted texts. These revisions were drawn from Jerome's own observations of the sky. It is a triumph: he paints himself as a hands-on astronomer, a trustworthy guide who has done more than just read the texts of the ancient sages. Two centuries later, the great Tycho Brahe will cite Jerome's works with respect. Having laid out his scientific credentials, Jerome then explains how science can improve on traditional astrology. And having done that, he ensures its popularity by going on to offer ten celebrity horoscopes.

Some were bigger celebrities than others, it has to be said. His father, Fazio, for example, was on the lesser-known side. But he also drew up a chart for Suleiman the Magnificent; for the Holy Roman Emperor Charles V; for King Francis I of France; and — perhaps most importantly — for Pope Paul III, himself a great believer in astrology.

This isn't just about catching public interest. It is a calculated bid for influence. Jerome fully believes he is worth listening to. He believes himself possessed of gifts and insights that he wants to share widely. He is approaching his fourth decade and keen to get on in the world. And so he dedicates the book to an old acquaintance of whom we'll hear more very soon: Filippo Archinto.

At that time, Archinto was the governor of Rome. In return for Jerome's dedication, Archinto told Jerome that Pope Paul might be open to a gift or two, which could establish the astrologer as a major public figure. So Jerome went all-out and created a horoscope of Christ to be presented to the Pope. And nearly forty years later, in 1570, that is what has landed him here, awaiting trial at the Inquisition and convinced that God has been protecting him through a guardian angel.

It seems odd that Jerome would think of me as any kind of useful guardian — he is in prison, after all. And this angel certainly did nothing to protect him from the machinations of Nicolo Tartaglia.

Chapter 2

Nicolo Tartaglia's story begins in the north of Italy, in the late fifteenth century. In the city of Brescia, fifty miles east of Milan, a baby boy has been abandoned in a haycart. The baby is taken to a home for waifs run by local nuns and they name him Michele. The boy grows, but not much: Michele is so small he becomes known as Micheletto.

Eventually, the sisters find Micheletto a kitchen job with a local dignitary. He is clearly resourceful: he manages to have himself transferred from the kitchen to the stables, and then, somehow — no one knows how — reaches the almost miraculous position where he owns a horse. Within a short time, he is known as Micheletto the Rider. Very quickly, the horse proves its worth when it enables Micheletto to become the local postman. He is the most reliable man to carry letters around the lakes, hills, and valleys of the Lombardy region of northern Italy, riding at good speed between Brescia, Verona, and Bergamo.

One of Micheletto's journeys takes him to the house owned by a gentleman of Verona, where he encounters a serving girl called

Maria. Evidently, Micheletto and Maria hit it off, because in 1496 they are married. In a short while, they have three children. It is the middle child, Nicolo, that is to become Jerome's nemesis.

These are not happy times in Italy, but Maria and Micheletto are happy enough for their part. Tragically, however, their love is to become the victim of political intrigue.

Ten years into their marriage, the world around them is falling apart. The French control the Lombardy region, but rumours abound that the Brescians are beginning to stir against the occupying force. Keen to nip the rebellion in the bud, France's commander for the region, General Gaston de Foix, orders that letters from Brescia be intercepted so that any plans for an uprising can be discovered. The postman's days are numbered.

Shortly after the order is given, Micheletto is ambushed by Gaston's men in the foothills of the Lombardy Alps. They find nothing incriminating: Micheletto's mail sacks contain only a handful of dull missives on trade agreements and the everyday correspondence of Brescian nobles. But the French soldiers kill him anyway.

Maria is devastated. Not only is her heart broken, but her household is stripped of its income and the family is thrown into poverty. And then, just when it looks like things can't get any worse, they do.

General Gaston's suspicions about the Brescians were well founded. By the time Nicolo is twelve, the simmering discontent has boiled over. In an act of supreme courage, the Brescian community of weavers and blacksmiths rises up and drives out the French soldiers stationed in the city's garrison. But not for long.

Gaston returns to Brescia with his entire army, and shows no mercy. In seven days of slaughter and arson, the French kill 46,000 Brescians, most of the city's inhabitants. The survivors — including Maria and her family — seek sanctuary in the cathedral. However, the French have no respect for the claim of sanctuary. Gaston's men burst in, brandishing swords and knives, and set upon the women and children. Though a screaming Maria does her best to protect her family, Nicolo takes an appalling blow to the face that cuts through his lips and palate and takes out most of his teeth.

Without Maria's care, Nicolo would have died. She has no money for professional medical help, but she has seen dogs lick their wounds and decides to follow that instinct. She bathes her son's cuts in her own saliva and keeps his injuries as clean as is possible in the aftermath of such brutal slaughter. Nicolo is hideously disfigured and robbed of the ability to speak. But he recovers. Once his strength has returned, a ruthless ambition kicks in. It is first manifest in his education. His schooling had stopped years earlier, immediately after his father's death, because there was no money. Aged fourteen, however, Nicolo arranges fifteen days of schooling on credit. This is only enough time for Nicolo to learn to form the letters of the alphabet up to k and he is thrown out once it becomes clear to the tutor that the pupil can no longer pay. Not, however, before the student has stolen his teacher's textbook. Returning home with his loot, Nicolo sets about instructing himself in the basics of literacy and numeracy. Then he learns everything else he can: 'I continued to labour by myself over the works of dead men, accompanied only by the daughter of poverty that is called industry,' is how Tartaglia describes that self-education.

When he is old enough, Nicolo covers his scars with as much beard as he can grow. He teaches himself to communicate as well as his wounds will permit. His associates come to know him as 'Tartaglia': The Stammerer. Thanks to Micheletto's illegitimate origins, he has no proper surname and he takes his new moniker with a defiant pride. It suits the young man to wear his grim brush with death as a badge of honour. Nicolo Tartaglia is not easily overcome.

ψ

In Milan, fifty miles to the west, Jerome is afforded opportunities which The Stammerer will never have. Jerome's parents are learning to live together under the same roof — they even marry, eventually — and his father teaches him Latin for an hour every morning. This is the language of scholars and debate, the language of learning, which no one teaches Tartaglia. That is why Tartaglia's writings will not be granted space in the academic libraries; The Stammerer will feel forever inferior because he can read and write only in the clumsy Venetian dialect of his youth.

Yet as Jerome learns Latin, he also learns about his culture. He studies horoscopes, unicorns, the rules of dice games, the features of the natural world, and more. He is taken to the houses of the great and good of the Italian Renaissance. He even remarks upon his father's elevated contacts in his journal: 'My father's reputation as a scholar was such that he was consulted by superior persons,' the twelve-year-old writes (rather pompously). He develops an appreciation for music and takes every opportunity to surround himself with musicians and singers. The shadow that comes with being born out of wedlock hangs over him, but does not seem to rob him of respectability.

For all his father's attentions, however, Jerome only attends university because of his mother.

It is 1519. Fazio is seventy-four, and every year seems to weigh heavier upon his stooped shoulders. Jerome is now eighteen and keen to fly free, but Fazio wants — deserves, the old man thinks — to keep his son at hand as a porter and scribe. Eventually, he reasons, Jerome can train as a lawyer, following in his father's footsteps. Fazio can even arrange for his son to inherit his annual one hundred crown stipend from the city courts. It is not a large sum, but it is something on which the boy could build a business. These are austere times and it makes perfect sense. Unfortunately, the son is not sensible. Jerome is not motivated by money and the desire for life's comforts. A life as a small-town lawyer is no path to greatness and that is the path he intends to walk. And so Jerome refuses to listen to his father's pleading. He wants to study medicine at his father's alma mater in nearby Pavia instead. Aware of the rift opening up between the men of her household, Chiara jumps in. Perhaps she wants her son to be happy or perhaps she is simply keen for Jerome to supplement the household income. Whatever the reason, she implores Fazio to let Jerome strike out on his own, leave Milan, and study at Pavia. They throw the idea back and forth, each hoping to score a palpable hit. And then, one afternoon, matters come to a head. In the middle of an argument over their son's future, Fazio hits Chiara across the shoulders with his staff. She falls, knocking her head on the table before she collapses onto the stone floor.

Unfortunately for Fazio, Chiara's sister, Margarita, is living with them. She witnesses the assault and begins to curse Jerome and his father, threatening criminal proceedings against her sister's assailant. Chiara, not terribly wounded and ever wily, sees her chance. She wipes the blood from her forehead and extracts a promise from Fazio: if she brings no charges, will he relent, and send their Jerome to the university to study medicine?

This moment of violence and calculated manipulation defines Jerome's path. Fazio has no choice but to concede and the youth enrols at Pavia in the medical school. Finding that he needs money to pay his way, however, Jerome takes up gambling. It is for this reason that he becomes the first person to apply mathematics to the problem of winning at cards and dice. His notes, which lay out what we now know as the mathematical rules of probability, give him a reliable means of beating his opponents. Erudite as they might seem with five hundred years of hindsight, those probability calculations did not begin as a noble venture in pure mathematics. They were simply Jerome's best chance of paying the bills.

ψ

Night has fallen. It may be the same night; it may be another. All I know is that I am here and Jerome is aware that I am sitting on the straw in the corner of his cell. There is moonlight and every now and then I see him look up from his writing and turn his head towards me. He no longer looks at all disturbed by my presence; he seems to welcome it.

'Tell me something,' I say. 'Did you exaggerate the story about Senator Lezun? Did he really fish you out of the canal after you had attacked him?'

The hint of a smile moves across Jerome's lips, but he says nothing.

'I bet you weren't really wearing armour,' I say.

His weak eyes light up at the word. 'I accept your bet,' he says. 'What are the stakes?'

I hesitate. Is it foolish to gamble against the inventor of probability?

Chapter 3

When it comes to the rules of chance, Jerome Cardano hasn't received the credit he deserves. The best-known version of the theory of probability's genesis begins almost a century after Jerome's death, with the confusion of a seventeenth-century gambler, the Chevalier de Méré.

Known to his friends as Antoine Gombaud, the Chevalier was a successful accumulator of other people's money. Back then, he made a good living by betting with less insightful men that he could throw a six in just four separate die rolls. He gave them even odds and pocketed their cash night after night.

Then he became ambitious. Gombaud's first problem arose when he extended his hustle to throwing a double six in twenty-four dice throws. He reasoned that it should work and tried it out. To his surprise, it was an utter failure. Gombaud took his disappointment, with furrowed brow, to an amateur mathematician called Pierre de Carvaci. He couldn't understand why it failed, either. De Carvaci then called on Blaise Pascal, the physicist and mathematician, for help. Also confounded, Pascal in turn decided to pass the problem to a Toulouse-based lawyer, Pierre de Fermat, of Last Theorem fame. Pascal and Fermat had a few back and forths on the subject in 1654, and eventually teased out the fact that it would take twenty-five throws to achieve a double six.

That this set of exchanges is widely considered the first effort into understanding probabilities would have horrified Jerome, with his ever-present desire for lasting fame. After all, he had first addressed the problem in a set of notes written more than a hundred years earlier, while still a twenty-year-old medical student in Pavia.

$$\psi$$

By 1520, according to Giovanni Targio, a tutor at the University of Pavia, Jerome is spending his evenings 'drinking and gambling in the taverns'. Targio writes of his student that he 'gave offence too often for his life to be free of enemies'. And the truth is that the young Jerome would often roam the streets of Pavia at night in a crude disguise, sword drawn — in flagrant contravention of the laws concerning naked weapons on public streets — 'determined to live some fantasy life', as his long-suffering tutor put it. That said, Jerome was no fool, and adroitly applied his mathematical mind to working out winning strategies for the gambling table.

Gambling is, at heart, about the random appearance of a string of numbers. In Jerome's time, however, no one believed in randomness. If numbers appeared to be random, that was only due to a lack of information. The prevailing view was that God controlled the die and the result of a roll (or the appearance of a card — cards being merely another way of creating random numbers) was the province of the Almighty. If you were out of luck, you must somehow have offended the Divine.

'So God knows everything. But what determines lotteries? — is it from Him?' Jerome asks in *On Subtlety*. His answer is somewhat confused. 'Not at all, but from some inspiration. However, our law states that lotteries are introduced and controlled by God.' Hard to see it now, perhaps, but Jerome's *Book on Games of Chance*, published posthumously in 1663, is an audacious first attempt to — as the physicist Stephen Hawking put it four and a half centuries later — know the mind of God.

There are thirty-two chapters. In Chapter 14, the world receives the first attempt at a law of probability to be applied when the die is cast:

> *So there is one general rule, namely, that we should consider the whole circuit, and the number of those casts which represents in how many ways the favourable result can occur, and compare that number to the rest of the circuit, and according to that proportion should the mutual wagers be laid so that one may contend on equal terms.*

In modern terms, Jerome means consider all the possibilities (the 'whole circuit'), then think about the number of ways in which you might get the result you want. Then find the ratio of those two numbers (the 'proportion'). That tells you how the bets should be placed.

More than a hundred years were to pass before Gottfried Leibniz came up with the same formula in 1676:

If a situation can lead to different advantageous results
ruling out each other, the estimation of the expectation will
be the sum of the possible advantages for the set of all these
results, divided into the total number of results.

And it would be nearly another century before, in 1774, Pierre-Simon Laplace delivers what is today generally assumed to be the start of probability theory:

The probability of an event is the ratio of the number of cases
favorable to it, to the number of possible cases, when there is
nothing to make us believe that one case should occur rather
than any other, so that these cases are, for us, equally possible.

Bravo Jerome! Yet back in 1520 he did not stop at a simple definition of the probability of a single event. He also worked out the probabilities associated with repeated throws of the die. That isn't easy. You might struggle, for instance to work out the probabilities — and thus the right way to bet (and accept bets) — with two throws of a die.

If you can only win with a one or a two, what are the odds of your two throws both giving you a win? Here is Jerome's solution, where a one is an 'ace' and a two is a 'deuce':

Thus, in the case of one die, let the ace and the deuce be
favourable to us; we shall multiply 6, the number of faces,
into itself: the result is 36; and two multiplied into itself will
be 4; therefore the odds are 4 to 32, or, when inverted, 8 to 1.

In more modern terms, we start with the first roll and divide the number of winning outcomes by the total number of possible outcomes, that is, two divided by six, or one third. The same is true for the second roll, and so we multiply the two results together: one third times one third is one ninth. That is, there is one favourable outcome in nine throws. Put another way — Jerome's way — you lose eight times for every one time that you win. In other words, your opponent's odds are eight to one.

And then three throws:

If three throws are necessary, we shall multiply 3 times; thus, 6 multiplied into itself and then again into itself gives 216; and 2 multiplied into itself and again into 2, gives 8; take away 8 from 216: the result will be 208; and so the odds are 208 to 8, or 26 to 1. And if four throws are necessary, the numbers will be found by the same reasoning ...

From this, Jerome deduced a general rule: Cardano's Formula, which is a means of working out and understanding probabilities.

Nor did Jerome stop there. Instead, he went on to deduce what we now call the 'law of large numbers', which shows that large numbers of repetitions of a probabilistic process produce a predictable overcome. Jerome has, to all intents and purposes, pioneered statistics.

Imagine flipping a coin four times. You know you should expect two heads and two tails, but you wouldn't be surprised if you got three heads and one tail, or vice versa, or even four heads or four tails. We know that, in processes that depend on chance, flukes happen.

Now imagine flipping that coin a thousand times. If you ended up with one thousand heads, you would assume the coin was weighted. Flukes are reasonable only with small numbers. Even six hundred heads out of a thousand tosses would be suspicious because, with such a large number of coin tosses, we expect something close to a fifty-fifty ratio.

Jerome worked out that one thousand tosses of a fair coin, where the odds of a head is one half, should give five hundred heads. In other words, you take the number of repetitions and multiply it by the probability of a particular outcome. That gives you the rough number of times that that outcome should occur. If it doesn't, someone is probably cheating.

Sadly for Jerome, this law of large numbers has been attributed to a mathematician, Jacob Bernoulli, who only worked it out — calling it his 'golden theorem' — 150 years later. But at least Jerome reaped its benefits, for it is his hard-won understanding of how the odds play out that tells Jerome, during a visit to Venice a few years after his graduation, that something is amiss at the gambling table of Senator Thomas Lezun.

ψ

It is 8 September 1526, the day of the Blessed Virgin's birthday. Those of a more devout disposition have spent the day in Venice's fine churches, reciting the appropriate prayer:

Let it be thy glory, O Virgin who destroyest all heresies, to restore unity and peace once more to all the Christian people.

Jerome, though, is not at prayer. Virgins are not really his thing — especially virgin mothers. In a quarter of a century, in *On Subtlety*, he will even dare to write of the 'notorious faked feat, that children are born to women without sexual intercourse'. It may not have been his wisest move, given his eventual arrest by the Inquisition.

Today he is in the Senator's house, trying to win back the property he had previously lost to his host. He is also hoping to win a night with a beautiful prostitute, a stake that had been on the table the night before. As a twenty-five-year-old man suffering from debilitating impotence, that had been a particularly alluring bet. It may even have distracted him from watching the Senator's play more carefully. Yesterday was a bad day. Jerome lost his clothes and his rings, as well as his chance of curing his problem through the attentions of a beautiful woman. But things are different today. On the Virgin's birthday, he has regained his footing. He knows now why he has done so badly. After so many hands of cards, it has become clear that the law of large numbers is not setting the scores straight. After all, this law says that all the cards should come up equally often and that no outcomes should be particularly guessable. Barring a few minor anomalies, everything should therefore follow the laws of probability that Jerome worked out a few years ago. But the statistics are skewed. The best explanation is always the most obvious: the Senator has been cheating. Jerome's brain is working well today. He has worked out that the cards are marked, and how, and in a few well-judged hands he wins back his clothes and jewellery. He sends them home with his servant because he knows he is about to make a hurried exit and doesn't want to be encumbered. He carries on with the game and wins enough money that he needs to bag it up. He puts all but one bag into the pockets of his cloak.

His heart is racing, but he is determined to make his accusation. He takes a moment to weigh his options. The front door is locked and the Senator has two servants in the room. The house is festooned with weapons, not all of them decorative. He glances up. Two lances hang from the beamed ceiling within easy reach of the servants. This is a brave — maybe foolhardy — thing to do, but the adrenaline is pumping and rationality is receding from his brain. So he draws his dagger.

Foolish and extreme as it may seem, Jerome has done his calculations and is confident enough to strike. His blade slashes across the Senator's face. He throws a bag of money across the table to make amends for his violence. 'Your master's cards are marked,' he announces to the servants. 'Unlock the door and let me go, or I will kill you.' They look to their master, who is weighing the bag of money in his hand. Jerome waits for the Senator's response, his heart still beating wildly. He has gambled well. The Senator puts down the money bag, considers his own reputation, and, with a hand wiping the blood from his cheek, orders the door to be unlocked. Jerome is free to go.

Outside, Venice is in darkness. Jerome begins to doubt. What sentence, he wonders, might he receive for an attack upon a respected public figure? It would be wise, he decides, to lay low until he can be sure the Senator doesn't have revenge in mind.

He dons a cloak, and — apparently — some leather armour. He keeps his weapon to hand. He wanders through Venice, keeping to the shadows. If the Senator has reported him to the magistrate, the peace keepers will be out looking for him.

And then, after a few hours of careful shadow hopping, it happens. With his nervous glances to left and right, his peering through the dark, Jerome is not paying enough attention to the ground beneath his feet. Slipping on a wet board, he fails to regain his balance and falls into the freezing, stinking waters of a Venetian canal. His clothing makes it hard — too hard — to swim. He begins to think he might drown.

Then, above his splashing, he hears the sound of an oar slapping at water. Out of the darkness looms a boat. Jerome makes a grab for one of the oars. He catches it, and the attention of the man wielding it. Shouts go up, and the crew begin to pull him, slippery, dripping, and heavy as he is, out of the water. Eventually, after much effort, he is lying facedown on the deck. He turns himself over, looks up, and finds the owner of the boat staring down at him. He recognises the man immediately, despite the bandages that cover half his face. Of all possible rescuers, it is Senator Thomas Lezun who looks down upon him. Jerome's eyes widen and his mind begins a new calculation of probabilities. Will Lezun, afforded this opportunity by fickle fate, take his revenge?

$$\psi$$

To Jerome, everything happens for a reason. If the dice land a certain way, it is because the Prince of Fortune decrees it. If he is rescued from a canal by a man whom he earlier insulted, it is God's way of making a point.

To me, it is just coincidence. Coincidence makes fools of us all — we cannot help but read it as significant, but we are easily deluded. How many times have you heard someone declare that 'everything happens for a reason'? It doesn't. Not if quantum theory is to be believed.

One of the most famous quotes in all of modern science is Einstein's comment that 'I cannot believe that God plays dice with the universe.' This was his response to the contention that some of the material world's occurrences are not preceded by a cause. A single atom of a radioactive material, for example, will emit a particle at random. There is no way to predict when this will happen and there is no known trigger. You can observe lots of atoms emitting particles and use the law of large numbers to give you an average time for emission to happen. But that tells you nothing about what makes it happen in any one atom.

Here it is worth stressing that this is only true, as far as we know, on very small scales. If I use a billiard cue to hit a ball into a pocket and know all the angles and forces involved, I can use the laws of physics developed by Newton to predict the paths of both cue and ball. Nonetheless, if I take one atom from one of those balls and fire it at two suitably sized and suitably separated openings — rather like two billiard-table pockets placed next to one another — there is no way to tell into which pocket it will drop. This is another manifestation of the double slit experiment that haunts our imagination. After a few dozen perfect repetitions I will get the sense of the most probable outcome, but each individual atom seems to make up its own mind. In each event, the effect has no cause.

The same is true if I fire a particle of light — a photon — at a mirror. There is a small probability it will pass straight through, and a larger probability that it will be reflected. If I fire a million photons at the mirror, perhaps only three will go through unreflected. But there is nothing special about those three. It is simply that random chance has dictated that they are not reflected. It is another cause-free event, nothing more significant than an outcome of the laws of probability. This is just like you winning the lottery when you have bought no more tickets than anyone else. God, it turns out, does play dice.

There is an obvious question to confront here. Why are the smaller particles — atoms and photons, for example — subject to purely random outcomes and events, whereas billiard balls are not? Nobody knows. Something happens to make the events of our world, the 'macro' world, deterministic, predictable, not random. All we can say is that we do not follow the same rules as the 'micro' world of atomic and subatomic particles, where probability theory, the theory birthed by Jerome, is the only way to predict the future. There is an unsatisfactory dark space in our knowledge.

$$\psi$$

'So what did Lezun do?' I ask.

'He handed me a set of his own clothes,' Jerome says. 'I can still see his smirking face now.'

'And were you wearing armour?'

'Where did you read that I was?'

'In your autobiography, *De Vita Propria Liber: The Book of My Life.*'

He smirks, calculating quickly that he has the bet sewn up. 'Interesting,' he says. 'I haven't written it yet.'

Thanks to my sudden confusion, I forget to ask about the prostitute.

Scientifically problematic, isn't it, that I am conveying information from Jerome's future? This is one of the strongest arguments against time travel; if you can influence the past, chronology, cause and effect, and common sense can be broken down far too easily. But, as we have already discussed, quantum theory — and its experiments — have shown time and again that cause-and-effect are not fundamental attributes of the universe. In order to understand why that is, we have to go back to basics, to the origins of quantum theory. And that story begins before history.

Ever since our species' birth, humans have looked for correlations between phenomena in the heavens and on Earth as a means to progress and success. Correlations between planting seeds at a new moon and reaping a bumper harvest, for example. Or perhaps we linked the appearance of a comet and an unexpected defeat in battle, or a particular arrangement of the planets with the birth of a king.

To make sense of these links, we began to note them down on stone and clay tablets. Later, we used paper. Once we had permanent records, we began systematically analysing that data, mining them for patterns that would help us to predict future events. A small subset of people even began to try to make sense of how the universe itself works. That subset of people — we now call them scientists — eventually showed that the hypothesis that the motions of the stars and planets cause effects on Earth could not be supported by the data. So while the predictors of the future carried on with their practice — unsupported by evidence, but eagerly watched by the general public — the scientists intent on understanding how the universe works abandoned the skies and began to pull the world apart. This is what led us, eventually, to the quantum.

Imagine taking a clock apart piece by piece. You will see its constituent parts — the cogs and wheels — and gain a sense of what they do. But now imagine wanting to know how the cogs and wheels obtain their properties. It won't be long before you are investigating the properties of iron. Grind away at some iron and you'll eventually get down to the very fundamental building block of that element: an atom of iron.

The atom has always been controversial. A Greek scientist, Democritus, contended a few hundred years before the birth of Christ that the atom was as far as matter could be split while retaining its essential character. Atoms of salt, he said, are sharp; atoms of water are smooth; atoms of iron are solid and strong. For millennia, though, the existence of the atom was just speculation. Only in the twentieth century did scientists — led, as it happens, by Einstein — reach consensus that they do indeed exist.

Atoms, we now know, are the bricks of our physical environment. They vary in size and have various different properties.

There are atoms in the air you breathe. They flow in rivers and make up the oceans. Your body is built from them. However, they are not indivisible. Atoms, too, can be broken down, into the particles that we call electrons, protons, and neutrons. And all of these particles — isolated atoms included — can behave strangely, in ways that you and I cannot.

Somewhat surprisingly, we have the alchemists to thank for the eventual emergence of the quantum rules that govern atoms and subatomic particles. It was they who became obsessed by light being the 'great primary cause', as the nineteenth-century scientist Robert Hunt puts it in his 1854 book, *Researches On Light*.

Hunt traces the science back to Benvenuto Cellini, one of Jerome's contemporaries. Cellini was a celebrated goldsmith and jeweller — much of his work commissioned by popes and nobles — an 'eccentric and extraordinary genius', Hunt claims. His contemporaries might have been less kind. Cellini was undoubtedly a gifted craftsman, but 'eccentric' would have to cover various charges of sodomy (brought against him by men and women whom he had used in what was known as 'the Italian fashion'), and murder, and several hefty fines and prison sentences for his misdemeanours.

Cellini's *Treatise on Jewellery*, published two years before Jerome's arrest, contains a passage discussing observations of a 'carbuncle glowing like a coal with its own light'. When held up to the lamp, and then put in a darkened room, this stone would light the room.

The publication caused quite a stir among the alchemists and they soon got to work seeking out, and creating for themselves, more and more materials that would 'phosphoresce', emitting their own light. A Bolognese shoemaker called Vincenzo Cascariolo made the biggest breakthrough: his alchemical experiments culminated in the synthesis of the first artificial phosphorescent material. Cascariolo managed to roast sulphur-rich barium sulphate for long enough to create a golden glowing mineral.

Eventually (and much more slowly than most like to think) the alchemists themselves transmuted into scientists, who began to investigate the properties of materials for their own sake, rather than in the pursuit of riches or the Elixir of Life. As Hunt put it, 'the hypothesis of the Alchymist has been converted into a probable theory by the discoveries of the modern chemist.'

Now the names become a little more familiar. Robert Boyle and David Brewster, for example, entered the fray, as did Newton, an alchemist through and through who wondered aloud whether solid bodies and light are interconvertible, and whether light is the source of the 'activity' of solid bodies. Eventually, it was realised that this light — all light — is a manifestation of energy. We know this energy as radiation and have more modern scientists such as Marie Curie to thank for these discoveries and their applications.

Radiation is not just light, though. There also exists invisible radiation, as Henri Becquerel discovered through his investigations of uranium salts. As the nineteenth century turned into the twentieth, we discovered a range of energetic particles that come from the component parts of atoms. Experiments showed that the atoms within these stones and metals break apart spontaneously, forming into different kinds of atoms and emitting energetic radiation. And as a result of examining this radiation very closely and precisely, we learned something very odd.

Have you ever seen an iron horseshoe in a blacksmith's fire? If so, you will know that, left in the heat, it glows red, then orange, then eventually white. This radiation comes from the individual atoms, with each one giving out a tiny bit of light that adds to the glow. The white glow is a composition of lots of different colours, just as you can obtain different colours from mixing different coloured dyes. Cellini, as a jeweller, would have seen the rainbow of colours projected onto a nearby wall when white light hits a diamond (a phenomenon, as it happens, that Jerome once discussed with King Edward VI of England). The process in the blacksmith's forge is, essentially, the reverse: all the colours united together as the heat takes hold.

As the twentieth century dawned, observations through spectrometers showed exactly how much of each colour of light an object like a white-hot horseshoe would give out. Scientists set themselves to explaining this distribution of energy in terms of the atoms' behaviour. It proved remarkably difficult. In the end, they only managed to solve it using the same technique that brings success in dice games: probability.

Imagine you need to roll a twelve, with two dice, and your opponent needs to roll an eight. You know there is only one roll that allows you to win: two sixes. But your opponent can win with any of five rolls: two fours, or a two and a six, or a three and a five on either of the dice. He is therefore five times as likely to win as you.

Max Planck, a German physicist working at the Friedrich-Wilhelms-Universität in Berlin, did the same calculation with atoms.

If you have ten atoms together, as if they were ten dice, certain colours of radiation are more common. That, Planck surmised, is because more combinations of those atoms' activity lead to those particular colours. Some colours are almost never emitted, much as a sixty is almost never achieved by throwing ten dice.

To match the observations of the colours emitted by hot objects, Planck divided up the energy of atoms into discrete amounts as if each 'quantum' of energy was a dot on a die. So there is nothing between 1 and 2 lumps of energy, just as there is nothing between 1 and 2 on a die. He then laid out the probability of the various possible quantum changes in energy. He didn't accept straight away that that was how atoms really were. He just wanted to see if it worked.

To Planck's surprise (and distaste, it has to be said) it did, and he discovered a peculiar relationship. According to the wave theory of light, we can quantify its colour in terms of 'frequency' (f). Imagine it coming towards you as a sequence of up and down oscillations: the frequency is the number of peaks in the wave that reach your eyes every second. Blue light has more peaks per second than red light; green is somewhere in the middle. Blue is therefore a higher frequency of visible light. Violet — and ultraviolet — is higher still.

Planck divided the amount of energy (E) in a 'quantum' packet of light (adopting a term already in use for other small packets of material) by its frequency (f). When he did so, he would always obtain the same number. We now know that number as Planck's constant. It is usually written as h, and E equals h times f.

Planck's extraordinary observation — that radiation must

exist as indivisible quanta whose energy is related to the properties of its waves — was the start of 'quantum mechanics'. This theory, which has taken over science as our ultimate explanation of how the universe works, splits every observable quantity — not just energy, but momentum, position, and so on — into indivisible, separable amounts, like the dots on a die.

Why? The answer is both simple and utterly confounding. It does so for no better reason than it fits the observations of what happens when you throw a horseshoe into the fire.

Chapter 4

It is 1525. The thirty-one-year-old William Tyndale is putting the finishing touches on his English translation of the New Testament, a book that will outrage and excite the population of Europe in equal measure. Jerome, just twenty-four, is more prolific. He already has two books under his belt. In addition to his manual on winning games of chance, he has written a volume about the failings of the medical establishment. He calls it *On the Differing Opinions of Physicians*, though it will later be published under a more inflammatory title. In its pages, Jerome has exposed the fact that doctors are using entirely subjective means — whims, mostly — to diagnose illnesses and prescribe cures.

Neither of Jerome's books are yet published, however, and if he wants to practise medicine he needs to keep it that way. The path to becoming a doctor has been difficult enough already. When he first attended the university, Jerome discovered himself to be a quick student and completed his Bachelor of Arts — an essential precursor to a medical degree — with ease. He took the examination for his medical conversion after just three terms of tutoring; most students needed nine. His insights into the profession were sharp enough that he had written *On the Differing Opinions of Physicians* even before his conversion course. Now, in 1525, the doctors here in Milan don't yet know about the book. Nonetheless, they will be keen to exclude him from their number.

Jerome is almost resigned to his fate. He has already experienced a hard time at his own university, despite everyone there knowing that his medical skills and understanding were beyond question. Thanks to an outbreak of plague, the Pavia faculty had decamped some two hundred miles to the east, to Padua. It was there, in the Great Hall of the University of Padua, that Jerome faced his final hurdle before qualification. This rigorous questioning and debate was a public event, with great potential for humiliation. However, Jerome's defence of his medical licence was so sublime that the faculty gave him a standing ovation. It was only when he withdrew from the Hall, so that the faculty could vote on his inclusion, that the knives came out.

The ballot was returned at forty-seven to nine against him and he was denied the right to practice medicine. He was outraged. 'No doubt the disgrace attaching to my bastardy, the odium of my attendance so frequently in places where dice and cards were the idols, and my rudeness in dispute, were more than these sages of the faculty could digest,' he fumed.

Jerome clearly appealed because two more ballots of the Faculty were subsequently taken. On the third, he finally won and was sent out from the medical school with the words, 'Go, and heal those who need you.'

The very next morning, sick of the prejudices of Pavia, he sets out to ride the two hundred miles to Milan. The journey is hard and Jerome is in the darkest of moods, consumed by a growing depression. Despite his triumphant graduation, his academic achievements cannot counter a self-loathing bred of an addiction to pornography, varied success in his sexual encounters with male and female lovers, and a gnawing problem with impotence. He is convinced he will never marry. The five-day journey through the foothills of the Alps sees him sink into despair. He is dogged by thoughts of suicide, a means to 'vanquish all ills for ever, and life with them'. And all this before his arrival in a city that will reject him out of hand.

Here is Jerome's problem, in a nutshell: his Pavia licence is not enough. In order to practise as a physician in his home city of Milan, the Milanese College of Physicians must also grant Jerome a licence. So on arrival he goes — the biretta of a physician on his head and the ring of his profession on his finger — to present himself to the College. There, he hands over letters of introduction from the faculty at Padua. When the Milanese faculty discuss his application, one of the higher-minded academics points out that the College statutes require that all members must be born within holy marriage. The rule had been flouted before, but it would not be flouted now. Jerome's application is refused.

$$\psi$$

'What did you do?'

Jerome is reclining on the straw mattress. I am sitting on the stool behind his desk. A rat scurries across the corner of the cell.

I dread his answer. Jerome has a penchant for the dramatic, especially where despair is involved. In *On Subtlety* he writes that the darker things in life always outdo the lighter moments. 'We come to grief and are disabled in an hour — indeed in a moment,' he says. 'To get enriched, to be healed, takes a long time. A vast evil arrives by chance, in a word; we are hardly kept safe by many friends and by their assistance. To sum up: so many calamities, griefs, pain, disease, dishonour, unrequited love, fear, poverty — and just one single good thing: not to be in need.' He goes on: 'What pleasure can be matched in magnitude to torture, what hope to fear, what happiness to bereavement, what agreeableness to imprisonment, what health to disease, or what wealth to the burdens of poverty, what honour to contempt or derision? Finally, there is that notorious last of all things, death, and even the joys of a thousand lives cannot be compared to its contemplation.' He is not a man easily inclined to happiness.

'I walked out of there without saying a word,' he says.

Dawn is coming and the light from the window casts a shadow of his beaky profile on the wall behind his bed. 'I went back to a room my mother had set aside for me in her house and lay down on the bed in black despair.' He turns his head so that he can look at me. 'Much as I am lying here now.'

You see my point?

ψ

As if things weren't miserable enough, in Milan Jerome finds himself living in a brothel. Fazio is dead and his money and properties are tied up in legal wrangles. Chiara, in desperate need of an income, has set herself up as a madam. She is born to it and the house is busy — too busy for Jerome's liking. No doubt he is reminded hourly of his impotence. He stays a few months, but is looking for an opportunity to get out. So when a doctor friend from Padua writes to tell him of the lack of physicians in Sacco, a small town a few miles southeast of Padua, he wastes no time. Anticipating a decent wage, he makes the journey and uses his scant resources, accumulated from gambling and tutoring, to buy himself a small house. Contrary to his friend's assurances, however, there is no medical work to be had in Sacco. Now aged thirty, Jerome is about to have his first experience of severe poverty. First, though, he is to fall in love.

Jerome meets his future father-in-law, Aldobello Bandarini, before he meets his future wife. Bandarini is hairy, dark-skinned, and rotund, with a personality to match his figure. An innkeeper, a businessman, and the captain of the Sacco garrison, he is a man of wealth, stature, and status. Bandarini also lives life boldly, hosting lavish and noisy parties, and standing drinks in his tavern. This is where the two men's paths cross, with Jerome attempting to earn a crust through his gambling skills.

Bandarini clearly likes Jerome. When he learns that the doctor's insomnia makes sleep impossible, Bandarini gives him licence to roam the town at night. Ordinary people would be challenged and perhaps arrested for nocturnal ramblings, but Bandarini tells his guards to let Jerome wander where he likes. Then, thanks to a house fire that turfs the Bandarini family out from under their roof and into a rented property next door to Jerome, he provides a beautiful young wife. It is lust at first sight. Jerome claims he saw Lucia Bandarini in a dream even before he set eyes on her in the flesh, and she excites something deep within him. According to his autobiography, the sight of Lucia Bandarini, his 'new-found love', immediately set him 'free from the bonds of impotency'. The poor girl has no idea of the troubles coming her way.

It almost didn't happen because Jerome initially decides he is too poor to marry. He dramatises the moment. "'Oh," said I, "what have I to do with this maiden? If I a pauper, marry a wife who has nothing save a troop of dependent brothers and sisters, I'm done for! I can scarcely pay my expenses as it is!'" Astonishingly, he contemplates an abduction, or seduction, but in the end there is no need. Her father is bewilderingly overjoyed at the match and offers Jerome not only a huge dowry but financial support for the couple for as long as they remain in Sacco.

It is now February 1532. The forty-one-year-old King Henry VIII of England is moving heaven and Earth, risking everything to be able to marry Anne Boleyn, the woman with whom he has fallen head over heels in love. Jerome, a decade younger than Henry, has had love and his future security handed to him on a silver platter, and he is moving heaven and Earth to make his own life as difficult as possible. Jerome is too proud to accept Bandarini's offer. He will marry Lucia, he says, but will not take a dowry and they will not live in Sacco. He will set out to make a proper living, and support his new wife himself.

ψ

'Why?' My stomach is tight, my jaw clenched.

'Why?'

'You could have been happy in Sacco. Lucia would have had her family and friends around her.'

Jerome's watery grey eyes narrow then soften. 'I was young,' he says. 'Young and in love, and foolish, and proud, and full of optimism.'

He is right. His writings make clear how wonderful and beautiful, how precious Lucia was to him. Which only makes his treatment of her all the more perplexing.

'You made her life so miserable.'

'Yes,' he says. He stares down at the flagstone floor. 'I know.'

ψ

It is one thing to choose poverty. It is entirely another thing to choose it for your pregnant wife. Jerome and Bandarini argue endlessly about the dowry and the offer of support, but the more educated man prevails in his stupidity. The newlyweds move to Milan, where the College of Physicians take great pleasure in denying him a licence for the second time. Chiara is happy and prosperous in her brothel, but Jerome and Lucia are unable — or unwilling — to take advantage of her earnings. Jerome describes the couple's lodgings as 'desolate'. He accepts only the minimum of his mother's charity, just enough to keep them alive. Their poverty brings ill health and, for Lucia, miscarriage. Also miscarried are Jerome's plans to become a published author. He has developed his writings about probability into a manuscript for the *Book on Games of Chance*. His book about the failings of the medical profession remains unpublished. He has also written a treatise on chiromancy, the study of portents read from hands. He offers them all around, but no one wants to publish them.

In an attempt to earn some coins, Jerome puts on his cloak, raises its hood over his face, and goes out into the night to practise medicine illegally. His only clients are poor and miserable themselves, and can pay little. He returns to gambling to supplement this meagre income, but in his present state he is poor company and unable to find victims willing to put up with his moods and outbursts. He goes — to all intents and purposes — mad. He has a permanent ringing in his ears; he rants in the streets against the College of Physicians; and he moans to his few patients and pupils about his illnesses and misfortune. So appalling is his situation that he seeks solace in the supernatural. 'In those days I was sickened so to the heart that I would visit diviners and wizards so that some solution might be found to my manifold troubles,' he writes. 'Many of them advised me, as it might be that on certain days I should drink only from cups that contained ochre, that on other certain days I should walk only on the left side of the arcades and shield myself from the moon's rays, or that on waking I should sneeze three times and knock on wood.'

All of this he does — and it doesn't help. 'Though I was at great pains to follow these advices they availed me nothing.' He looks at the stars and the planets. He examines the lines in his hands. There is no relief from his fate.

Destitute, weak, and tempestuous, Jerome decides that the Cardanos must move to the countryside to subsist more cheaply. In

April 1533, with Lucia pregnant again, they leave Milan and head northwest to the small country town of Gallarate. Their possessions are tied to the backs of the only two mules Jerome can afford to hire. He and his wife make the twenty-five mile journey on foot. It is too much for Lucia, and she suffers her second miscarriage.

'So,' I say. 'Your decision to refuse your father-in-law's generosity reduced your beautiful young wife to a life of poverty, two failed pregnancies, and no prospect of a child. And how did Gallarate work out for you?'

Jerome's gaze is fixed on the window. 'I gambled away my wife's jewels and our marital bed.'

'And then?'

'And then, when all else failed, I became a writer.' He raises himself up on one elbow, turning to face me. He says, with one eyebrow raised, 'Worse. A science writer.'

This is the moment when I understand that Jerome is starting to know me.

Somewhere in Gallarate, almost certainly in an inn and within the fellowship of the dice, Jerome has met a local nobleman. Filippo Archinto — not yet the Governor of Rome to whom Jerome will dedicate his second book on astrology — is intelligent and curious, a lover of life. People instinctively like and trust him, and he does not abuse their trust; he is known to be kind, as well as good company. Perhaps it was Archinto who won Lucia's jewels and maybe the bed. He certainly feels some kind of obligation towards the destitute couple: learning that they have nowhere to live, he allows them to use a suite of rooms in his summer house.

Archinto does more than just put a roof over the Cardanos' heads: he also creates a job for Jerome. There is no medical practice to be had, and the town is too small for Jerome's skill at calculating odds to serve him for long; once a few have lost money, word quickly gets around and opponents vanish like mist in spring sunshine. Archinto is impressed by Jerome's intellect, though. He has academic pretensions himself, which he channels into work as an amateur astronomer. Although he can make observations, he finds himself unable to write them down in a coherent manner. Jerome, he realises, can turn his thoughts into dancing prose that will impress the scholars of Europe.

And it does. To Archinto, Jerome is a 'genius'. He has, Archinto writes in his letters, 'dwelt upon the secrets of the skies'; he 'knows of numbers as much as any man; delights in music and performs it with grace; can spend upon philosophy great argument; and in the little time when he might be idle puts down notations and sketches of many ingenious engines and toys'. It is all true — but none of that pays. Not like ghost writing for Archinto pays. Constructing Archinto's book, *Judgements of the Astronomers*, keeps Jerome and Lucia alive with housing and a small income.

These are happy days. Chiara takes time off from the brothel and comes to visit. She finds her son and daughter-in-law in good health and good spirits. Optimism fills the air and Chiara writes back to Milan that Jerome is about to become famous. 'I am certain,' she says.

She is wrong. Out of nowhere, in the middle of one winter night, Archinto disappears. When the morning comes, Jerome looks for him, hoping to pick up the discussion on writing that they were enjoying last night. Archinto is nowhere to be found. After a couple of hours, Archinto's lawyer arrives and tells Jerome what has happened. A messenger came during the night with a sealed command from the Holy League. Barbarossa, Grand Admiral of the Ottoman fleet, has attacked the coast of southern Italy. Archinto has been conscripted to fight for the Pope's army, with orders to leave immediately. He is obedient and faithful; he has gone. For Jerome and Lucia, it is a disaster.

The lawyer quickly makes his way to the summer house. Everything is to be closed up during the owner's absence, he says. Since his client left no instructions about the Cardanos, they are to leave immediately.

Once again, the Cardanos are homeless. Once again, Lucia is pregnant.

Chapter 5

It is 1534. England is recovering from its disappointment at the birth of a daughter, Princess Elizabeth, to King Henry VIII and Anne Boleyn. Lucia Cardano, though, will be happy with any living child, whatever its sex. This time she is lucky. Despite the circumstances, her third pregnancy progresses without incident.

Jerome and Lucia are now living in a rented hovel in Gallarate. They grow a few vegetables and forage for food in the countryside. The mathematician is gambling again, though without much success. Yet Jerome is optimistic — perhaps insanely so. 'I hear continual voices and dream visions promising fame,' he announces to his mother in a letter. He is even welcoming the prospect of an extra mouth to feed, convinced that the third time is the charm. 'The trinity of Lucia's pregnancies will be fruitful, for there is a sacred quality in this number,' he tells Chiara.

Ten days later, on Thursday 14 May, he is proved right — though not without a scare. The birth itself is uneventful, but shortly afterwards Lucia begins to haemorrhage. Finally, Jerome gets to practice some worthwhile medicine: he saves his wife by administering a coagulant. The baby is a boy, a weak, mewling infant with 'small, white, restless eyes' as Jerome recalls. The child is deaf in his right ear and has a curved spine. The third and fourth toes of his left foot are webbed together. Hindsight grants Jerome more perspective than he was capable of at the time. 'Had I known then of the wickedness that was to seize the boy and the evil that he was to cast upon other people,' he writes later, 'I might in my anguish have been tempted to cast him aside and let him pule in vain.'

Of course they don't cast the baby aside. Given the child's weakness, however, they resolve to baptise him sooner rather than later. Facially, he looks something like his paternal grandfather and Jerome is tempted by the name Fazio. Lucia objects and prevails. After some argument, on a warm, sunny Sunday, he is anointed Giovanni Battista.

The baptism takes place in Lucia's bedchamber, but it does not go well. The scene could have inspired the Brothers Grimm to concoct the story of Sleeping Beauty: at the last minute, just as the newly named child is lifted from the font, an unwelcome visitor arrives.

'Suddenly a great wasp flew into the room, though it was by no means the season for wasps, and circled angrily around the child,' Jerome recalls in his autobiography. 'We beat it away in fear that it would harm him with its poison, and it flew angrily against the wall and for some moments continued a resounding noise like a drum. The creature disappeared as suddenly as it had come, no one seeing its departure although all eyes were upon it.'

Jerome is left shaken and fearful. 'I was filled with horror by the menacing premonition that a spell was cast upon my son by an evil insect, and that though its sting had not afflicted him it had through the powers of darkness poisoned his spirit.'

Later in life, Jerome was to advise that the real power of curses lies in the accursed person's self-sabotage. 'It is of use to withhold belief,' he writes in *On Subtlety*. Now, though, the curse takes hold. While Lucia is still recovering from the haemorrhage, Jerome falls ill. He suffers a high fever and 'agues and vomitings'. He is so weak that he is unable to work for a month and the household falls into penury. His solution is as stubborn as the mules that brought them to Gallarate: they will move back to Milan.

It is late August 1534. Poverty mandates that they have no mules this time, just a flat trolley laden with books that Jerome pulls behind him. Lucia carries her baby. For the three days and nights of the journey, they beg for food and sleep in fields.

On arrival in Milan, they are unable to find anywhere to live — Jerome's mother, it seems, is no longer a resident. For four weeks, Jerome, Lucia, and Giovanni spend their days and nights on the streets. They are still begging for their food; now, in the city, they are considered vagrants, and are frequently abused and spat upon by passers-by. Jerome's obstinate, self-aggrandising decisions have reduced the beautiful, radiant Lucia to a life of utter misery and shame.

In October, as the temperature begins to drop, Jerome gives up the attempt to provide for his family through begging and applies for them to enter the workhouse. His application is successful and, from October until the end of December, Jerome and Lucia perform menial tasks in exchange for shelter and meagre rations of food. Then, as the year draws to a close, a fellow recipient of workhouse charity pulls Jerome into a dark corner and whispers in his ear. The man is old, decrepit, and clothed in rags. 'You are cursed,' he tells the astonished Jerome. 'If you want, I can help.'

The old man, it turns out, is a wizard.

ψ

Jerome won't say whether he consulted his wife. Nor will he be drawn on the issue of why a decent wizard should be so down on his luck as to be living in the workhouse. He avoids dwelling on the legalities of the activity; performing spells is not an uncommon practise, but it is witchcraft and the church is operating a zero-tolerance policy on magic these days: witches and wizards are burned at the stake if caught. Nonetheless, Jerome, ever prone to going with his gut, always game for a new experience, and still convinced that a bright future lies just around the next corner, agrees to let the old man help.

The wizard's prescriptions are many and various. The first task is to rid Jerome of the demon that has taken possession of the good doctor. A daemonifuge herb performs its purgative functions, then the wizard pummels Jerome with stones to grind the bones of the devil within. The ringing in Jerome's ears, the wizard tells him, is the screams of that devil. 'From that time forth I never experienced that ringing again,' Jerome wrote in *On Wisdom*.

Now things take an even stranger turn. The wizard knows of a stagnant pool, not far from the workhouse. It will, he says, provide the perfect stage for the next part of the exorcism.

They venture to the pool on a night when the full moon is reflected in the dead centre of the water. 'Toads leapt upon its edge and the plants and weeds on its surface were of a glutinous kind with slime,' Jerome recalls. On the wizard's instruction, Jerome puts his head into the water exactly where the moon's reflection is sat and keeps it immersed while the wizard utters an incantation. He can hear only a murmur of the wizard's voice, but the words, he says, cause the vapours of his curse to be 'drawn upward to the atmosphere of that nearest and most benign of the planets.'

I think he means the moon. Anyway, the incantation is done and Jerome must now dry his head upon unfouled straw and give the wizard a pouch of powdered alicorn — that is, powdered unicorn horn. He will not say how he came by the means to purchase this, but he clearly considers it a worthwhile investment. 'From that time forward, this strange necromancer told me, I should fare famously. And I have no cause to declare him false.'

Nor we. Within a few weeks, Jerome is on the way up. A sceptic would argue that it is pure coincidence that Filippo Archinto happens to return just now from the Holy War. The fact is, Archinto is shocked to learn — from a letter written by the wretched Lucia — that his friend is living in the workhouse. He immediately travels to Milan and provides Jerome and his family with lodgings in the city. What's more, he nominates his 'genius' friend for a new position as a public lecturer on geometry, astronomy, and arithmetic.

The post has been created in the will of a philanthropic resident of Milan called Thomas Plat, who left a small legacy as a salary.

Archinto's social status is on the rise, thanks to his noble birth and service to the church, and he uses his growing influence to secure Jerome the job. The prestige of the position — and the fact that Jerome cannily makes his lectures more popular by replacing the more boring parts of geometry with geography, and arithmetic with architecture — opens up work as a tutor. In 1535, Jerome earns fifty crowns, which provides his household with enough money for a mule, two servants, and a nurse for the baby. His mother comes to live with them, bringing extra income from her brothel. A year after leaving the workhouse, the Cardanos are respected members of Milanese society. So respected, in fact, that Jerome decides to have another shot at obtaining a medical licence from the College of Physicians.

He fails. There are some strings that even Archinto can't pull.

'And so you decided to get your revenge by publishing *Bad Medicine*?'

'My book was called *On the Differing Opinions of Physicians*,' Jerome says. The smile playing on his lips betrays his acceptance of my appraisal — yes, revenge, in part. 'I was practising in secret, treating a whole host of patients who told me that the College's physicians had done nothing for them,' he says. 'It seemed to me that these people deserved to know how little we doctors agreed on the best way to treat a patient.'

'Ah,' I say. 'So it was a noble gesture. Not revenge?' I smile. 'Not even a little bit?'

Jerome lies back down on the straw, his face lost in the dark again. 'Maybe a little,' he mutters into the gloom. 'But I had so little to lose, and as it turned out, so much to gain.'

ψ

It can't have hurt that Jerome's oldest friend from university had inherited a printworks. Ottaviano Scoto owed him a favour too: years ago, when they were both students in Pavia, Jerome had loaned him some essays. Scoto, a pale young man burdened by a sense that he was Jerome's intellectual inferior, had compounded his feelings of inadequacy by losing the pages. Now that Scoto is the owner of a printing press, Jerome thinks it might be time to call in the debt. The response he receives overwhelms him with joy. Scoto's reverence for Jerome's intellectual abilities is still in full bloom: he can't believe that Jerome's writings have yet to be published. He volunteers for the honour and privilege — and is happy to do so unpaid. 'Joyfully he said that he would take all the risk of publication,' Jerome says, 'and that even if he were to lose by it, it would be small cost for the privilege of being the first to put Doctor Cardano into the vision of the public.'

And that is how *On The Differing Opinions of Physicians* came to public attention. With a publisher's eye for profitable controversy, Scoto puts it out as *On The Bad Practice of Medicine in Common Use*. It is an immediate hit and the Milanese College of Physicians is furious. Doctors, according to Jerome's text, are apt to prescribe unnecessary medicines and to bleed and starve patients to their detriment. He discusses 'the tribal insecurities of men who banded themselves together and showed to the world a surface of pomp and learning that satisfactorily concealed from the beholders the depth of ignorance beneath.'

A vicious, but entirely predictable, backlash hits Jerome with full force. The faculty broadcasts the fact that Jerome has no practice at all, and is therefore in no position to criticise. This book, they say, will put patients' lives in danger. He is nothing more than a lecturer on arithmetic, with a famous grudge against the College. What's more, they point out, the text is almost incomprehensible.

In this last point, at least, they are right. In his hurry to publish, Scoto has failed to perform any proofreading. There are more than three hundred misprints and typographical errors. 'I reaped nothing but shame,' Jerome said later. 'The book damaged me in every respect save one: it sold in great quantities … My friend Scoto rejoiced, but I grieved.'

In the end, he needn't have worried, for he is about to become a recognised and fully employed doctor, even without the College's approval. To the faculty's astonishment and outrage, Jerome has received a job offer: would he like to become official physician to the city's Augustinian friars?

Once again, the hand of Filippo Archinto is deeply involved. Francisco Gaddi, the city's Augustinian prior, has been ill for two and a half years with a skin disease — almost certainly a type of leprosy — that none of the city's doctors can cure. After two years, depressed and burdened by a morose fascination with death, Gaddi asks his friend Archinto's opinion. You must, Archinto says, call for my friend Jerome.

And so he does. Quickly realising that the city's doctors have tried every pharmaceutical available, Jerome quizzes the prior at length about his diet and routine. Shocked at the self-neglect of the spiritual life, he prescribes some pampering. Mortification of the flesh is out: no more sackcloth; no more proscription of soap and water so that the outer dirt mirrored the inner man; no more fasting and sleep deprivation. Gaddi was to sleep well and eat well, denying himself neither fish nor wine.

Now, six months later, Gaddi is a changed man, in rude health and grateful for it. And Jerome has a smile on his own face. All those nights shuffling through the streets of Milan in cloak and cowl, working for the recompense of coppers and owed favours, risking criminal charges if he were to be caught practising illegally — they have not been in vain. Justice is finally paying her debt. He accepts the job at the priory gleefully; he does not need the permission of the Milanese College of Physicians to practise there. Though the position is not well paid, it is a long overdue poke in the eye for the College physicians. In Jerome's eyes, that is payment enough.

ψ

'So you came good,' I suggest to Jerome with a smile. 'You put Lucia and Giovanni in a decent home, finally published a book, and became a respected member of Milanese society. And you made the first recorded prescription of a spa treatment.'

Jerome sits up. He looks unimpressed. 'A lot of good it did Gaddi. He died in some prince's dungeon ten years later,' he says. 'How do you cure a man whose chief ailment is a tendency to make enemies of the whole world?' He pronounces the question loudly, like an actor on a stage. Then he hesitates, and looks around his cell. His demeanour collapses, as if he has suddenly realised his end might be no better. 'I should have seen the signs,' he says. 'There must have been signs.'

'You don't believe in signs.'

Jerome looks at me as if I have gone mad. 'There are always signs,' he says. 'You just have to know where to look.'

'Do you know this sign?' I ask. I get to my feet and scrawl on the rough cell wall using a small pebble. My sign looks like a three-candled menorah; the Greek letter *psi*. Jerome stares at it, one eyebrow raised in curiosity.

$$\psi$$

'I do,' he says. He pauses, and points at my drawing. 'What does it mean to you? Is it part of your magic?'

I throw down the pebble and laugh. 'You could say that.'

Psi — Ψ — means everything to a quantum physicist. It is the symbol assigned to the full mathematical description of an object. That object might be a universe, or an electron. No one knows if *psi* is a real thing, with physical properties that you can measure in experiments, or just a mathematical summary. It is known as the wave function.

Remember Planck's frequency, f, which describes the colour of light emitted by a hot horseshoe? That's actually a measure of how fast a back-and-forth or up-and-down oscillation occurs, and it is related to the wavelength of the light. Imagine a child on a rope swing. If the rope is long, it swings back and forth slowly. A short length of rope creates a fast oscillation. Waves are the same. Long wavelength gives low frequency; short wavelength gives high frequency. This wave idea is how an Englishman, a Frenchman, a Swiss, and a Dane arrived at the theory that gave birth to *psi*.

First, the Englishman. This is Ernest Rutherford, who in 1908 instructed his students to fire radiation at thin sheets of gold to see how the radiation would be affected by the gold atoms. The way the radiation was deflected showed Rutherford that almost all the mass of an atom is concentrated in a tiny region at its centre. This is what we call the nucleus and it carries a positive electrical charge. In a much bigger space around the nucleus — if the nucleus were the size of a fly, the atom would extend to the size of a cathedral — there is a balancing negative electrical charge: the electrons. Taken as a whole, the atom has no electrical charge.

Enter the Dane, Niels Bohr. In 1913, he suggested that, following the model of the planets going round the sun, the electrons existed in particular orbits around the nucleus. When they gained energy — as when an iron horseshoe is put into the fire — they moved out to orbits that were more distant from the nucleus. Eventually, they would fall back to their original orbits, giving the extra energy back as emitted radiation: that is, light.

Bohr came up with a formula that would explain why the light energy only took on certain values — Planck's discrete, quantum values — and nothing in between. The electron orbits, he suggested, must exist at particular distances from the nucleus, with no available orbits in between them. That would mean the electrons had to make jumps involving a particular amount of energy, which corresponded to Planck's quanta.

Now we come to a quick cameo by the Swiss: a patent examiner called Albert Einstein. He showed in his special theory of relativity, published in 1905, that there is a speed limit to the universe. The mathematics describing how radiation is emitted by objects only makes sense if nothing can move faster than the speed which we now call the speed of light.

So here comes our final player: a French prince called Louis de Broglie (it is pronounced '*de Broy*'). De Broglie was a graduate in medieval history, as it happens, but also a keen student of the new physics. He was the genius who suggested in 1924 that Bohr's electron orbit, Planck's frequency (and its corresponding wavelength), and Einstein's universal speed limit work together to explain the origins of quantum theory.

An electron orbiting a nucleus, he said, could be thought of as a wave that travels round in a circle. For the atom to be stable, the distance it travels in that circular orbit must correspond to a certain number of complete waves, or 'wavelengths'. That would make the wave perfectly smooth and continuous, with no ugly jumps as the circle is completed.

That is only possible if the orbits are at particular distances from the nucleus. Imagine the electron is at a distance from the nucleus where its circular orbit allows it to complete, say, two whole waves in a trip round the nucleus. Now move the electron out, radially away from the nucleus. Eventually you will reach a point where the circular orbit would be able to fit three complete waves. In between those two positions, de Broglie said, is forbidden ground.

De Broglie worked out the mathematics and showed that we

can indeed treat the electron as either a wave, or a particle. The numbers all add up. That means you can talk about its energy, or you can talk about its frequency and wavelength. Either is fine; both are correct.

All of that works fine without Einstein's input, in fact. But to make everything work, de Broglie's electron waves travelled too fast: their speed exceeded the speed of light. That was forbidden by Einstein's relativity. To get round this, de Broglie reasoned that there must be a component of these waves that represents something outside the physical universe ruled by Einstein's speed limit.

'Ah,' says Jerome, 'It lives in the *aevum*.'

I should have seen that coming. During the course of his lifetime, Jerome gave a great deal of thought to time and space, and our place within it. He eventually concocted his own theory concerning the geometry and topology of the universe, which suggests a new, inaccessible dimension.

He called it *aevum*.

Aevum is the place where intelligences dwell — where information resides, effectively. It has a temporal dimension: it is eternity, and is 'analagous to the centre of the sphere,' Jerome says in *On Subtlety*. 'The centre corresponds to every point of the circumference and remains stationary as the sphere rotates. In this way, eternity remains fixed within the infinity of time. It does not expand, it does not flow, it is always at rest.' Our visible cosmos resides within this *aevum*. 'The universe, apparently at rest, is contained within eternity, and within the universe time flows.'

'Am I right?' Jerome is beaming, clearly pleased with himself.

'Honestly,' I say. 'I don't understand your *aevum*. And I don't really understand the consequences of de Broglie's work properly, either.'

The part of de Broglie's wave that exists outside of our physical universe is found within its 'phase'. Phase is an abstract concept that has long been associated with waves, or anything that performs a repeated motion. It is often used to compare the state of two or more systems: two swimmers could be described as 'in phase' if their arm movements are synchronised. Put two grandfather clocks next to each other, and the motions of their pendula are 'in phase' if they move to left and right at the same time and 'out of phase' if their movements are always in opposing directions. In a single system, such as a travelling wave, the phase is a quality that describes the stage of the wave: the peaks and troughs of an ocean wave, for example. Quantifying it, we say the phase describes the proportion of each oscillation that has passed.

But de Broglie's genius was to make phase into something physical, a quality in its own right, and locate it outside of our normal physical reality. An electron has a mass and a velocity, say, *and* it has a phase. No one knew what it meant, exactly, but it made the mathematics work. The numbers added up.

Thanks to de Broglie, we finally had a convincing explanation for the glow of radiation from a hot horseshoe: quanta of radiation whose properties were ultimately determined by a phase that transcends the physical universe. That, however, may have been the last fully convincing moment in quantum theory.

$$\psi$$

'And the phase is *psi*?' Jerome has hardly batted an eyelid. He looked confused by the concept of electrical charge and raised an eyebrow at the mention of the planets going round the sun. Aside from his interjection about the *aevum*, however, the rest was absorbed without murmur. I can't tell if he understood it.

'No. The phase is just one component of *psi*.' I hesitate. 'There's another part to this whole thing. Do you remember the numbers you discovered as a solution to the cubic equation? The square roots of negative numbers?'

He scowls. 'Ah, the impossible quantities,' he says. 'What useless sophistry.'

I smile. 'Not useless at all. When I learned about them at school, they were called *imaginary* numbers, but they're not imaginary and they are certainly not useless. They make up an essential part of the quantum phase.'

'They are taught to children?' Jerome looks sceptical.

'Yes. Although they are largely unwelcome in young minds. Children always ask what the point of them is, if they're imaginary. The point is, they aren't imaginary. It's just a bad label. They are used in all kinds of ways.' I walk over to the wall, pick up a pebble and scratch an *i* onto the wall. 'When you put your imaginary numbers together with your work on chance, you can derive quantum theory. And that can explain how everything in the universe works.' I turn back from the wall, expecting to find him looking triumphant, or at least pleased. I have forgotten, just for a moment, that this is a man who will find a fly in every ointment.

'Can it explain why I was arrested?' he says, lowering himself onto the mattress. 'Can it explain on what charges I am being held here?'

The sceptical question, the refusal to celebrate his achievement, draws the air out from within me in a deep sigh. I sit down on the straw beside him.

'You don't need imaginary numbers or probability to explain that,' I say. 'Although they both played their part, as I understand it. Mostly, though, your stay here' — I wave a hand at the cell — 'is the work of Nicolo Tartaglia.'

Chapter 6

1535, the year Jerome and his family leave the workhouse and begin to rise in Milanese society, is the year William Tyndale is arrested in Antwerp for heresy. All across Europe, religious disputes are boiling over. Sir Thomas More is soon to be executed for refusing to recognise King Henry VIII as head of the Church of England. In Holland, the Anabaptists attempt to take control of Amsterdam — and fail. In Germany, they succeed, ruling the city of Münster for five months. Six French Protestants are burned in front of Notre Dame Cathedral for distributing printed condemnations of the Catholic Church. They are only the thin end of the wedge, so King Francis I has decided to ban use of the printing press.

None of this bothers Nicolo Tartaglia, who is just trying to keep his head down and earn a living. It hasn't been easy. Perhaps through circumstances, perhaps through bad luck, The Stammerer is not blessed with an amiable personality. Since growing up to become an intelligent, scholarly young man, he has been driven out of his hometown of Brescia not once, but twice for being avaricious, morose, and, above all, rude to everybody.

Biting his thumb at the Brescians, he has settled in Venice, where he writes down ideas in his own crude version of the Venetian dialect. Though it is far from the erudite language of scholarship, these ideas are clever and innovative enough to pass muster. A local nobleman has recognised his talent and helped him with money to study at the University of Padua. Tartaglia has become a teacher of Euclid's theorems. It is a solid enough career for any self-taught man and one that Tartaglia is pleased to call his own. No wonder, then, that he is so incensed when Antonio Maria Fior arrives in town, his eye firmly fixed on Tartaglia's hard-won job.

Fior is an inferior mathematician with a superior mathematical lineage. His teacher was a celebrated scholar, Scipio Ferreus, a professor at Bologna. And here Fior was in the right place at the right time. In 1505, Ferreus worked out a way to solve a particular type of equation known as a cubic. Cubic equations are those that include x raised to the power of three. That is, x multiplied by x multiplied by x. A simple way to visualise it is as the volume of a cube, whose sides are of length x. The volume is its length multiplied by its breadth multiplied by its height: x^3. If x is two, say, the cubic power is $2 \times 2 \times 2 = 8$.

In Jerome's time, however, there is no mathematical notation, so all problems are laid out in convoluted sentences such as '*cubus p. 6. rebus aequalis 20*'. In modern notation, we would write that as $x^3 + 6x = 20$. This makes it significantly easier to solve. The idea is to find a numerical value for x, but at this point some cubic equations were thought to be impossible to solve.

In 1494, the Franciscan friar Luca Pacioli had published a book, expansively illustrated by Leonardo da Vinci, in which he explained why there would never be a general solution to the cubic equation. Ferreus didn't have a general solution, but he could solve an equation of the form $x^3 + bx = c$. It was an impressive innovation, involving many complicated routines of substituting one variable for another, and making a variety of assumptions that allow a solution to emerge. However, it was still not the long-sought general solution. Give Ferreus something like $3x^3 + 8x^2 - 7x = 15$ and he would be undone. A new solution was necessary.

To a medieval mathematician, a new solution to an equation is a precious thing. It is a stick with which to beat your enemies, a dagger by which to slay them and take their jobs. That's because no one publishes their solutions then waits for their peers to give them acclaim. The modus operandi is to humiliate your rivals by challenging them to mathematical duels.

The public relish these occasions. This is Renaissance sport, featuring intellectual heavyweights sparring in the mathematical ring. It isn't just about mathematics, either. In February 1498, for instance, Leonardo da Vinci joined a public battle with various mathematicians, astronomers, astrologers, and physicians. This was a contest designed to establish the pecking order of all the arts and sciences. It's not clear who won the battle, but the occasion did lead da Vinci to write that painting — because it is based on observation, mathematics, and geometry — should be considered a science. 'No human investigation may claim to be a true science,' he says, 'and if you would say that those sciences which begin and end in the mind possess truth, this is not conceded, but denied for many reasons. The foremost [reason] is that such mental discourses do not involve experience, and nothing renders certainty of itself without experience.'

Battles between mathematicians, though, bring entire populations to the city square. This is about more than who can make an argument for their intellectual superiority. Here, there are concrete problems to solve, points to score, answers that are definitively right, wrong, or unachievable. There is cheering and catcalling, a referee, champions, and underdogs — all accompanied, of course, by drinks and snacks. Once the contest is over, the winner takes all. In theory, the loser is supposed to buy his conqueror a dinner for each of the unsolved problems. In reality, he often has to cede fame, money, and sometimes even a job. No wonder Tartaglia is worried.

A battle starts with the competitors giving each other thirty problems to solve. They must be able to solve their own puzzles, so anyone with a solution that others don't have is at a distinct advantage — they can set problems that their rival simply cannot figure out in the allotted time.

It is not clear whether Ferreus ever used his solution to the cubic equation in a public mathematical battle. But he knew its value and only passed it forward from his deathbed. The recipients were Fior, his pupil, and Ferreus's son-in-law, Annibale della Nave. The son-in-law did nothing with it, but Fior had no qualms about putting his teacher's skill to work.

Fior was a lazy student and an unskilled mathematician. Ferreus's solution was his only card and he knew it. For two decades, he guarded it jealously and used it to great advantage in public duels. But then, in challenging Tartaglia, he pushed his luck too far. Fior wanted Tartaglia's job and Tartaglia needed it to live. So when Fior challenged The Stammerer to an algebraic duel, Tartaglia took it as seriously as an assault on his life. What Fior didn't know was that Tartaglia had already been working on cubic equations. Moreover, unlike Fior, he was born to the task. In many ways, Fior's rising fame had worked against him. Tartaglia had discussed Fior's wielding of cubic equations with a Brescian schoolteacher called Zuanne da Coi, a tall, lean, hollow-eyed man, with a strange, slow gait. Da Coi had then encouraged Tartaglia to think about the possible solutions and so he had — with some success. Tartaglia now possessed a solution for equations of the type $x^3 + ax^2 = b$, and $x^3 = ax^2 + c$.

Following the usual protocol, Tartaglia and Fior exchanged problems two weeks prior to the duel. Tartaglia evidently knew — perhaps from da Coi — about Fior's weak grasp of mathematics and exploited it by setting him a range of mathematical tasks, including some that involved the new cubic solutions.

Predictably, Fior's problems were only of the type for which he alone had a cast-iron schema that would generate solutions. The Stammerer took them to bed with him and rattled the equations around in his mind. It worked. Almost certainly because he had already worked out solutions to equations of a similar kind, Tartaglia worked out Ferreus's method a full eight nights before the contest was to take place. He then went on to work out the solution to yet another form of the equation. By 13 February 1535, the night before the public duel, Tartaglia knew he would win.

Tartaglia was impressive. Fior's face must have worn a look of pure horror when, in less than two hours, Tartaglia solved all the problems that had made Fior's living for the last thirty years. Fior, for his part, couldn't solve any of the equations Tartaglia had set. There was a clear winner. Fior asked Tartaglia to demonstrate his methods, but The Stammerer walked away. He didn't even claim his thirty dinners. For him, a burnished reputation was enough.

ψ

'Tartaglia was good, wasn't he?'

The sun is almost up. Jerome's skinny student, Rudolf Silvestri, has just delivered a bowl of boiled vegetables, setting it down on the floor at my feet. When he stood up and seemed to look right through me, I took offence, until I remembered that I am not sure whether I am actually, physically present. Jerome has picked up the bowl and set it on the table. Now he is fixing it with an intense stare, as if it contains the secrets of the universe.

'Fior was hardly a great opponent,' he says, eventually.

I smirk at Jerome's withering snark, but decide not to pursue the point. 'So, at this point, Tartaglia is riding high in mathematics and you are riding high in medicine, thanks to the job at the Priory. What went wrong?'

'It wasn't enough, I suppose,' Jerome says into the bowl. He picks out a piece of turnip and brings it to his lips. 'I didn't want to just cure the sick. I wanted to move things forward — to understand.'

'Understand what?'

'Everything.' He takes a bite of the turnip. It is soft and he barely chews it before swallowing. 'Life. The cosmos. Everything. This is my calling, my only task.'

In this regard, Jerome reminds me of Erwin Schrödinger, the most famous quantum mechanic of them all. Schrödinger once said that to understand ourselves is the only task of science.

Fittingly, it is Jerome's mathematical innovations that made Schrödinger's work possible.

Remember the glowing horseshoe? The radiation comes from atoms of iron. More precisely, it comes from an electron that surrounds the nucleus of those atoms. As the electron loses energy, it effectively falls between these different orbits. The properties of that radiation are determined by the wave that fits onto the orbit. But that wave is no ordinary wave: it is *psi*, the wave function.

The wave function has a phase, which — as de Broglie pointed out — must exist outside of the pure physical reality of the time and space that we know. And the only way to fully describe that phase is to use mathematics that involve the imaginary numbers, written as *i*.

i is the square root of -1. I know I haven't explained exactly what Jerome did to discover it yet (and that discovery is almost a decade away still), but trust me for now: the fact that Jerome did anything with such an oddity is extraordinary. In his day, negative numbers were themselves suspect. People understood what it meant to have two apples. They did not understand (and perhaps we don't, really) what *minus* two apples could possibly be. How, then, to take a square root of a negative number? After all, you know that the opposite process — squaring a number — never gives a negative? We know that 2 x 2 = 4. We know that -2 x -2 = 4. We know that the square root of four is therefore two and minus two. But what is the square root of minus four?

Where we call such solutions 'imaginary', Jerome termed them 'sophistry'. If you came across imaginary numbers at school, you probably hated them. Who knew they were hated for centuries prior?

If mathematical conundrums like this leave you floundering, you are in good company. They left Jerome floundering, too. But he was the first to face up to them. When the geometer Heron of Alexandria was working out the volume of a truncated pyramid in the first century AD, he found he had to deal with the square root of minus sixty-three. Heron just fudged his calculation, quietly dropping the negative number. In 250 AD, the Greek mathematician Diophantus derived a square root of a negative number in his calculations. He assumed he had made a mistake. That was also Jerome's first thought when he discovered he had to deal with the square root of a negative number. But he checked, and there was no mistake. He gave the notion a name: 'impossible quantities'. And then he went on to put them to use in solving fiendishly difficult equations. It was, he said, 'impossible' to find numbers that solved these equations, but there are certain 'objects' — his 'impossible quantities' — that get the job done.

Translators argue about the proper way to interpret Jerome's Latin in this passage of his 1545 book on the art of algebra and arithmetic. Some believe he regarded the imaginary numbers as 'mental tortures' that had to be ignored before you could complete the calculation. Others infer that he was simply saying that the 'imaginary parts' become lost in the process, making all things work out in the end. One translation contends that Jerome said this mathematics is 'truly sophisticated', another that he thinks it 'truly imaginary'. Still others have translated it as 'truly sophistic' — that is, something of an elegant fudge. And yet we see him demonstrating a difference between 'pure negative' numbers (minus two, for example) and 'sophistic negative' numbers, such as the square root of minus two:

Note that $\sqrt{9}$ is either +3 or -3, for a plus [times a plus] or a minus times a minus yields a plus. Therefore, $\sqrt{-9}$ is neither +3 or -3 but is some recondite third sort of thing.

That 'recondite third sort of thing' is central to our very existence, as it turns out. It is not only used in engineering bridges, aeroplanes, and mobile phones (and truncated pyramids, of course), but it is central to the most fundamental theory of physics. Without Jerome's imaginary numbers we cannot explain how the universe works or even the most basic processes of atomic physics. It was Schrödinger, an Austrian, who worked that out over the Christmas holiday of 1925.

Schrödinger is the Pablo Picasso of physics. He was a genius, who represented reality in a way that defied all previous conventions and was also unrepentantly uninterested in living a 'conventional' life. Schrödinger's lifestyle gave no quarter to the morals of his contemporaries and he left tragedy and a trail of shocked bystanders in his wake. Early in his professional life he became maths tutor to a thirteen-year-old girl called Itha Junger. In his diaries he admits he groomed Itha by writing her poems and engaging her in grown-up conversations about religion and science. Their maths lessons included petting, cuddling, and fondling her until she was seventeen, when he finally felt able to seduce her. Their secret relationship continued on and off for years. In 1932, she fell pregnant and sought an abortion because Schrödinger would not leave Anny, his wife. The procedure went badly and left Itha unable to have children.

Anny tolerated Schrödinger's insistence that their relationship be 'open', commenting that 'it would be easier to live with a canary than a racehorse, but I prefer the racehorse'. In 1934, Schrödinger took up a position at the University of Oxford and brought both his wife and a new mistress, setting up their household in a *ménage à trois* that the Oxford dons could barely tolerate — especially as his mistress was pregnant.

None of this detracts in any way from his genius — perhaps it even stimulated it. He was certainly stimulated at Christmas 1925, when he left his wife behind, picked up a different mistress — this one's identity remains unknown — and a copy of Louis de Broglie's work, and went on a trip to the Swiss Alps.

By the time he returned home, he had a world-changing equation that would win him a Nobel Prize. We now call it the Schrödinger equation, and we write it like this:

$$i\hbar \frac{\partial}{\partial t}\psi = H\psi$$

I have filled in the gaps on the wall with my stone scrawl.

Jerome shakes his head. 'I don't understand your notation,' he says. Of course he doesn't. The notation is a product of the centuries that are to follow.

'The *h* with a bar through it is Planck's constant divided by two times *pi*. Two things placed next to each other means multiply them together. *d* divided by *dt* means the rate of change of the thing it precedes.'

'*H*?'

'*H* is something called the Hamiltonian operator. It contains information about the energy in the system and varies from system to system — you just plug it into the formula.'

'Hamiltonian?'

'It's named after a William Hamilton.'

Jerome raises his eyebrows at the name. 'The Archbishop?' he says. There is hope in his voice.

'I'm afraid not. He was an Irish mathematician. Or he will be.'

Jerome looks disappointed. 'Perhaps a descendant, then?'

'Perhaps. William Hamilton had Scottish ancestors, I know that.'

It seems enough to satisfy him. 'And this Hamiltonian,' he points a finger at the wall. 'What does it do?'

Now there is a question.

Schrödinger's equation is a means to predict the physical properties of a system — not just an electron in an iron atom that might emit radiation, but anything that follows quantum laws. Because *psi* has a phase that exists outside of our ordinary three dimensions of space, we know that *psi* must itself inhabit another realm, something like Jerome's *aevum*. Physicists call it Hilbert space, after David Hilbert, the mathematician who came up with the concept. Hilbert space has an infinite number of dimensions. It is generally described as an 'abstract' space, rather than a physical one, but there is an ongoing argument as to whether the wave function is physical or abstract. We'll get to that. First we have to deal with a slightly more accepted oddity associated with the wave function's behaviour within Hilbert space: it relies on Jerome's probability theory.

Physicists use the Schrödinger equation to make predictions of what to expect from quantum experiments and processes. Very early on, they realised that — although the equation officially depends on the wave function, which describes the unfolding events — all we can actually do is work out the *square* of the wave function, Ψ^2. The square of *psi* gives you a probability associated with anything you want to know. And the right choice of Hamiltonian operator, which depends on the physical system you're working with, tunes the equation to each particular thing.

Understandably, Jerome looks confused. I don't know if I can make it any easier. '*Psi* squared is linked to the probability of the atom's position, say. You specify a position, and it will tell you how likely you are to find it in that position if you carry out a measurement.'

'So it is like a bet?'

'Worse. It's like a bet that depends on imaginary numbers.'

'And the equation won't guarantee that you win?'

'All it will do is give you the odds of winning.'

'So the atom might be somewhere else when you look for it.'

I take a deep breath. 'It's not somewhere else. It's everywhere else. And nowhere. Until you make a measurement.'

I stare at Jerome to see how he reacts to this. It baffles me. It baffles everyone I know.

Clearly, it doesn't baffle Jerome. He looks intrigued. And rather pleased.

'So you call it into existence with this measurement?' he says. 'Like a magus summoning a demon?'

My brow furrows. Is that right? I hesitate, then decide it's as good as anything I've heard. I nod, as sagely as I can.

$$\psi$$

This is not quite the whole truth. This understanding of what the Schrödinger equation tells us is known as the Copenhagen interpretation of quantum mechanics because it was invented and promulgated — quite forcefully, at times — by the great Dane, Niels Bohr. It is the most popular of the interpretations among physicists — but not the physicists who think particularly deeply because it is actually bankrupt as an explanation of what happens in the quantum world.

Niels Bohr is widely considered the father of quantum theory. His research centre in Copenhagen, which was sponsored by the Carlsberg brewery for a time, was certainly the hub for the theory's development. Bohr was well liked by most of his contemporaries. Although he was frustrating to argue with, since he gave no quarter and generally considered himself right about everything, he was far from dull. There were times when he engaged his colleagues in mock gun fights because he had developed a theory — drawn from watching American cowboy movies — that the person who draws second will always win.

For a highly intelligent man, Bohr was astoundingly slow to

understand anything going on around him. From movie plots to presentations at physics conferences, he tended to need everything explained to him several times from several different angles. That didn't prevent him from being a Nobel Prize–winning genius, however.

Here's how Bohr's explanation of the quantum world works. Go back to the double slit experiment, where a single photon has two apertures to pass through in order to reach a detector on the other side. Think of the photon as an arrow and the detector as two archery targets placed side by side. If the bullseyes are lined up with the two slits, you'd expect the arrows to hit only the bullseyes. That would be the everyday, 'classical' outcome. But after you watch the quantum archer fire one arrow at a time for a few hours, you go and look at the target. You find the arrows are clumped around a few particular points, with no arrows at the points between.

Now let's change the spectator's view. Instead of looking at the archer, look at the slits to see which one the arrow goes through. After a few hours, go and look at the targets. You'll now find that the arrows are only in the bullseyes. The act of watching the arrow pass through the slits has changed the outcome.

Going back to the photon in the double slit experiment, what does this mean? According to Bohr's Copenhagen interpretation, the photon — faced with two possibilities given by the two slits that it can pass through — exists as a wave. That means it is in a region of space, but not definable at one particular point, much like a wave on the sea. So we can't say anything about the photon's specific position. All we can do is talk about its wave function — the mathematical description that Schrödinger gives us. But remember, the wave function's only purpose is to describe what the outcome of a measurement might be. Its square gives the probabilities of the various outcomes occurring. It says nothing about the photon's life story, only its final outcome. In fact, Bohr would say there is no life story — there is no photon to talk about, effectively — until that measurement happens.

However ... nowhere in the Copenhagen interpretation is the 'measurement' defined. Instead, Copenhagenists talk about the 'collapse' of the wave function from a wide array — a 'superposition' — of many possible states into just one definite state.

So, what *is* the measurement? How does it cause this 'collapse'? There have been lots of attempts to find out. One group of researchers based in Vienna are working particularly hard at this and are convinced that it is something to do with information carried by the photon.

Think of the photon as Tartaglia's father, Micheletto, the postman. His job is to carry letters from one place to another. He doesn't know what is in the letters, but we do know they contain information. Whoever reads the information is carrying out a measurement. While the letters are sealed, only the sender knows what is in them. To everyone else, all possibilities are open, which is to say that they are in a superposition of all possible pieces of information. Maybe it's a bill? Maybe it's a love letter? Maybe it's an order for unicorn's horn? Once they are opened and read, however, the information is suddenly defined. That is a measurement.

What the Viennese researchers did was, effectively, to try melting the seal on those letters. At first just a little bit. They heated up the photons, and then they looked at the pattern on the detector. I know I'm conflating letters, photons, and arrows here, but what they saw was that the arrows were less clumped. The pattern was less stark. As they heated the photons more, so the distribution of the arrows became more and more like you'd expect in the everyday world — focussed on the bullseyes, with just a few strays elsewhere.

The act of heating, they suggested, tells us something about measurement. As anything gets hot, it radiates energy. That's how we began this whole tale, if you remember. And the different wavelengths — the colours of that light energy — can contain different quantities of information. The result of heating was that some of the information about the photon — the contents of the letter — became readable. It's as if the yielding wax allowed a corner of a page to roll up and be read. Only a few words became visible, but it was enough to allow the reader to see that it was probably a love letter. Perhaps the words 'your lips' came into view. Now it was definitely not a bill, nor an order for unicorn horn. It might still have been something other than a love letter. It could have been an instruction to keep silent: a threat, for instance, to a witness of a crime not to let a single word about the incident pass their lips. Whatever the truth, not all possibilities about the nature of that information are open any more. And so the weird quantum pattern is less stark. The Viennese researchers think that the information radiating away from the hotter photons is enough to read their location as they pass through the slits — in other words, that it could betray which slit the photon passed through.

This leaking of information is known as 'decoherence' and that's about as much as we understand about it. There seems to be, according to the Copenhagenists, a threshold beyond which information leakage will cause the wave function's superposition to 'collapse'.

Niels Bohr first presented what came to be known as the Copenhagen interpretation of quantum physics in 1927, at a conference in Como, Italy. His presentation came after what was almost a two year silence about quantum theory, during which he had wrestled with the idea that a photon was both a wave and a particle, and — simultaneously — neither. In the end, his view was that we could say nothing about what it actually *was*. All we could say is what we see. In that way, the observer becomes an integral part of the system.

'There is no quantum world,' he once said. 'There is only an abstract physical description.' And later, 'an independent reality in the ordinary physical sense can neither be ascribed to the phenomena nor to the agencies of observation'.

The reception to Bohr's talk was mixed, with most people rather underwhelmed. One of the physicists in the audience, Eugene Wigner, was entirely unimpressed by Bohr's Como lecture. It would change no one's mind about their own take on the meaning of quantum physics, he said. But, in the end, Wigner was wrong. Louis de Broglie, for instance, changed his mind, coming over to Bohr's point of view. There are many who wish he hadn't. But to understand why de Broglie came round, we have to pursue Jerome's story, too.

Chapter 7

Nicolo Tartaglia's self regard is dramatically inflated after his victory over Fior. He is aware he now possesses unique knowledge that will provide him with an income for years to come. He guards it jealously and to good effect. His enhanced reputation brings him various teaching posts, including a public lectureship in Milan as well as his Venetian teaching post. This gives him the money and time to turn his mind to subjects no one else is thinking about.

One of those subjects is artillery. No one has yet put mathematics to work in the service of kings and generals at war and Tartaglia finds himself well equipped. However, he is filled with a pacifist's remorse. For all his vain self-aggrandising nature, he remains a devout man of high moral standards. The creation and refinement of killing machines does not sit well with him, as this passage from his writings makes clear:

One day I fell to thinking it a blameworthy thing, to be condemned — cruel and deserving of no small punishment

by God — to study and improve such a damnable exercise,
destroyer of the human species, and especially Christians
in their continual wars ... I destroyed and burned all my
calculations and writings that bore on the subject. I much
regretted and blushed over the time I had spent on this.

That attitude only softened when Suleiman the Magnificent, Sultan of the Ottoman Empire, began to slaughter Christians in the crusades. 'It no longer appears permissible to me at present to keep these things hidden,' Tartaglia says. 'I have hence resolved to publish them partly in writing and partly by word of mouth, to every faithful Christian, so that each may be better fitted in offence as well as defence.' In 1537, convinced he can help the Pope's cause, he self-publishes his treatise on the trajectory of cannonballs, *The New Science of Artillery*.

Good as the book is, it misses its target. Tartaglia simply doesn't have the reach, the connections, to attract the attention of military minds. So when Jerome gets in touch two years after its publication and offers to help, Tartaglia takes the bait.

Jerome is desperate to know Tartaglia's secrets. Having heard a great deal about them, he is suffering a base professional jealousy. Jerome sees algebra as his specialist subject and yet he is painfully aware that he doesn't know all there is to know about it.

It is a particular problem because he has decided to write a revolutionary book on arithmetic. This is to be no high-minded endeavour to impress his academic rivals, but a maths book for the common people. It will be sixty-eight chapters long, with an introduction that lays out the journey the reader is to take.

That decision was made fourteen years ago and, in the years since, he has already completed much of the book. So far, *The Practice of Arithmetic* has dealt with basic operations, such as multiplying and dividing. It has taught integers and fractions, and even the supernatural properties of numbers. Jerome has reached the more advanced chapters now. The problem is that Tartaglia's confidante, the hollow-eyed Brescian schoolteacher Zuanne da Coi, has come to Milan. He is talking loudly about Tartaglia's new solutions to the cubic equations. Jerome knows these new solutions must be published in his book. But how to get them?

Fortunately, Jerome has trained a student, Lodovico Ferrari, whose talent will help achieve this tricky task. Lodovico came to Jerome's household as a fourteen-year-old servant on 30 November 1536. Initially, he was a replacement for his cousin, who had run away from the notion of working hard for a living. Unlike his cousin, Jerome noted, Lodovico could read and write, so he put the boy in charge of the household secretarial work. When Lodovico proved able and clever, Jerome began to teach him mathematics. Lodovico may well have been the guinea pig on whom Jerome tested the efficacy of his blossoming maths text.

Five years later, in 1541, Jerome had become convinced of Lodovico's gift and was determined to give it space to flower. He resigned one of his teaching posts, as a lecturer at the Piatti Foundation in Milan, knowing this would give Lodovico an opportunity to break into academia.

There were only two candidates for the newly vacant position: Lodovico and the teacher Zuanne da Coi. The choice was resolved in a public maths duel — of course — and Lodovico won easily.

The twenty-year-old servant is now also a lecturer in geometry, feels a great debt to Jerome, and is fiercely loyal to his cause.

Jerome and Lodovico work well together, researching new algebraic solutions to difficult problems. But still they cannot crack Tartaglia's secret. Eventually, they decide on a more direct approach. They will ask Tartaglia to share his knowledge for the common good.

Aware of Tartaglia's prickly reputation, Jerome and Lodovico enlist a local bookseller named Zuan Antonio as a go between. He will convey Jerome's carefully phrased question to Tartaglia and bring back the reply, with all communication directed to Antonio as a third party.

It starts off politely enough:

Master Nicolo Tartaglia, I have been directed to you by a worthy man, a physician of Milan, named Master Girolamo Cardano, who is a very great mathematician. And because he has understood that you have been engaged in disputation with Master Fior, putting to him for a wager certain questions that could only be answered by knowing the general rule for resolving the case of the cubic, which general rule you had found by your own discovery. Therefore his excellency prays you that you will kindly make known to him the rule discovered by you, and if you think fit will make it public under your name in his present work, but if you do not think fit that it should be published he will keep it secret.

The reply, predictably, came in the negative:

Tell his excellency that he must pardon me: when I propose to publish my invention, I will publish it in a work of my own, and not in the work of another man, so that his excellency must hold me excused.

The full back-and-forth becomes a little tedious — most particularly for those involved. However, Jerome does not give up. He even offers never to publish the solutions, if Tartaglia will only reveal them. Tartaglia refuses.

Jerome then asks for the list of questions which Fior had set him in the competition for Tartaglia's post. These Tartaglia reluctantly supplies, commenting that, 'his excellency, whatever his competence, will be unable to resolve them, for to do that would mean his excellency had a wit like to my own, which he has not'.

Chiding his correspondent for such an 'unhandsome reply', Jerome remarks that whatever Tartaglia might think, he is himself 'nearer to the valley than the mountain-top.' It is then that he drops into the conversation the fact that he has passed a copy of Tartaglia's book on artillery to a friend. A friend who happens to be Alphonso d'Avalos, Marquis del Guasto.

The Marquis, as Tartaglia knows — and Jerome knows that he knows — is a powerful military man. Jerome now dangles the chance of an offer of employment before his rival. He invites Tartaglia to come to Milan to meet the Marquis. Jerome's masterful move puts Tartaglia in a bind. In the margins of a letter from Jerome, Tartaglia scrawls out his frustration: 'I am reduced by this fellow to a strange pass, because if I do not go to Milan the lord marquis may take offence, and such offence might do me mischief, I go thither unwillingly: however, I will go.'

And go he does. When he arrives, however, the Marquis is out of town. Jerome invites Tartaglia to visit his house instead. There he repeats his promise that, if he could just learn the secrets of the cubic equations, he will refrain from publishing them. Tartaglia brazenly says he is not willing to take Jerome at his word. Jerome swears several oaths and says he'll even write the solutions down only in code, so that no one can discover them, even after his death.

And it is then — for reasons that remain unclear — that The Stammerer wavers. He says he will ride off in search of the Marquis; on his return, he will show Jerome the solutions. Seizing the moment, Jerome says no — tell me now. And, using a poem as a cover, Tartaglia does so:

When the cube and things together
Are equal to some discreet number,
Find two other numbers differing in this one.
Then you will keep this as a habit
That their product should always be equal
Exactly to the cube of a third of the things.
The remainder then as a general rule
Of their cube roots subtracted
Will be equal to your principal thing
In the second of these acts,
When the cube remains alone,
You will observe these other agreements:

You will at once divide the number into two parts
So that the one times the other produces clearly
The cube of the third of the things exactly.
Then of these two parts, as a habitual rule,
You will take the cube roots added together,
And this sum will be your thought.
The third of these calculations of ours
Is solved with the second if you take good care,
As in their nature they are almost matched.
These things I found, and not with sluggish steps,
In the year one thousand five hundred, four and thirty.
With foundations strong and sturdy
In the city girdled by the sea.

Yes, it seems unbelievably, painfully contrived. But this is Renaissance Italy; this is how they are. Indeed, Tartaglia is by no means the first algebraist to summarise his insight in poetic form. Decades earlier, for example, the mathematician Luca do Borgo had expressed his algebraic rules in Latin quatrains.

Tartaglia is proud of his rhyme: it 'speaks so clearly', he tells Jerome, 'that, without other example, I think your excellency will be able to understand the whole'. Jerome, in reply, assures Tartaglia that he has understood almost everything about the solution on his first reading of the poem. Remember your promise not to publish, is Tartaglia's response. And then there is silence. Five months pass. Tartaglia, regretting his weakness and hearing disturbing rumours, writes to Jerome from Venice:

I am very sorry that I have given you already so much as I have done, for I have been informed, by a person worthy of faith, that you are about to publish another algebraic work, and that you have gone boasting through Milan of having discovered some new rules in Algebra. But take notice, that if you break your faith with me, I certainly shall not break promise with you (for it is my custom); nay, even undertake to visit you with more than I have promised.

Jerome holds his nerve. He has indeed published a book containing algebraic rules, *The Practice of Arithmetic*. But, no, he hasn't broken his promise. He sends Tartaglia a copy, so that he can verify this for himself. Tartaglia's response is cruel and petty. He writes to Jerome that the book contains nothing new. It is simply a synthesis of what we already knew. What's more, he adds, the book is confused and full of mistakes: 'The whole of this work of yours is ridiculous and inaccurate, a performance which makes me tremble for your good name.' He points to one particular howler: 'your excellency has made such a gross mistake that I am amazed thereat, forasmuch as any man with half an eye must have seen it — indeed if you had not gone on to repeat it in divers examples, I should have set it down to a mistake of the printer'.

Nor has he finished. So numerous are the errors, Tartaglia says, and sometimes so rudimentary, that Jerome is clearly incapable of innovative work. The idea that Jerome could have worked out Tartaglia's solution to the cubic equation for himself, he writes, 'sets me off laughing'. Enter Lodovico Ferrari, Jerome's brilliant pupil, spoiling for a fight. Ferrari clearly believes it is below his beloved master's dignity to pursue the quarrel. But Ferrari has no trouble with brawls himself, having lost two fingers to a rival's dagger in a tavern fight. Now, seeming to relish taking on Tartaglia, he writes on his master's behalf:

You have the infamy to say that Cardano is ignorant in mathematics, and you call him uncultured and simple-minded, a man of low standing and coarse talk and other similar offending words too tedious to repeat. Since his excellency is prevented by the rank he holds, and because this matter concerns me personally since I am his creature, I have taken it upon myself to make known publicly your deceit and malice.

He even makes a mathematical joke. With all Tartaglia's talk of square roots, cube roots, and more, he says, 'I promise you that if it were up to me to reward you, taking example from the custom of Alexander, I would load you up so much with roots and radishes that you would never eat anything else in your life.'

You probably had to be there. But the reality is that very many people were, for the correspondence between Tartaglia and Ferrari is sold in the streets like newspapers. Indeed, it circulates around Europe as other academics take it upon themselves to explain to everyone what the dispute is about. The row has become famous, as have its protagonists. The unexpected side effect is that the public is learning mathematics in order to follow the soap opera. To be precise, the public is learning about cubic equations and conic sections.

That might seem implausible now, but in Renaissance Italy

the public is still aware that these subjects belong to a revered intellectual lineage. For most of history, conic sections — the points where a cone meets a surface — were just a mathematical challenge, the geometrical Sudoku of the professional mathematician. The Greeks, for instance, loved playing around with these puzzles and, a couple of hundred years before the birth of Christ, a geometer and astronomer called Apollonius of Perga wrote an eight volume treatise on the mathematics of the conic section. More than a thousand years later, Omar Khayyam, the Persian mathematician and poet, wrote his *Treatise on Demonstration of Problems of Algebra*, in which he showed that the intersection of a hyperbola with a circle provides a geometric method for solving cubic equations. This, it is worth saying, is not a pointless pursuit. Conic sections trace out an ellipse or parabola and can be used to calculate the trajectories of planets through the sky, or artillery through the air.

$$\psi$$

'I simply had to have them in the book.' Jerome's tone is defensive, almost sly. 'Students of algebra needed to know these solutions. Do you understand?'

I nod, but slowly and with my head inclined; I can see his point, but I can also see Tartaglia's. 'I suppose so,' I concede eventually. 'It's funny: we teach our children to solve these equations, but I don't think they ever know why they are useful.'

Jerome frowns at this. 'Why not?'

'Most of our children give up studying mathematics as soon as they are able.'

'And come back to it when they are older?'

I shake my head. 'No. They almost never come back. They may go through their lives knowing the formula for solving a quadratic equation, but never once apply it to anything.'

'So there are bakers on the streets of sixteenth-century Milan now who feel more in tune with mathematics than the educated children of your city?'

'More in tune than the educated adults, if I'm honest.'

You know it's true. Remember this? x equals minus b, plus or minus the square root of b squared minus four ac, all over two a. It is a phrase that means something to people educated to secondary level across the world. Few of us realise that the Babylonians knew it too, as the formula to solve a quadratic equation of the form (as we would write it now): $ax^2 + bx + c = 0$.

Jerome and Tartaglia and all their peers certainly knew this trivial piece of mathematics. What hadn't been resolved was how far you could go. Could there be solutions for all equations that involved the cubic: x^3? And what about the quartic: x^4? The quintic: x^5?

Years earlier, Scipio Ferreus had gone some of the way with cubic solutions and Tartaglia had gone further still. Jerome and Lodovico had derived two more solutions to algebraic problems. Jerome was eager to publish them, but he was hamstrung. His work relied on Tartaglia's solution, revealed in the poem, and he had promised not to publish that. And then he found a way to sidestep The Stammerer.

ψ

In 1543, someone — we don't know who — whispers in Jerome's ear that it was Scipio Ferreus, not Tartaglia, who came up with the first solution. If that is true, why not go straight to the source? Or, given that the source is dead, the source's son-in-law?

Here, fortune smiles on Jerome. It turns out that the son-in-law, Annibale della Nave, is still very much alive, and living in Bologna. Jerome and Lodovico pitch up at his door. Annibale shows them Ferreus's papers. In an 'aha' moment worthy of any great detective story, they see the solution — and see immediately that it is the solution that Tartaglia hinted at in his poem. Yes, Tartaglia might have worked out a solution to the cubic equation, but Ferreus had worked it out first. That means the solution is not the exclusive property of Tartaglia. And so, Jerome reasons, it could be published in his next work, a book on the 'Great Art' of algebra.

'My heart was skipping,' Jerome says. His eyes glint. 'After that, the solutions poured out.'

'The quartic? And all the various cubic solutions?'

He nods. 'And now I was free to publish them. My oath to Tartaglia was irrelevant'

'Because you could publish his solution as Ferreus's work.'

He grins: a wide, ugly leer. 'Exactly,' he says.

In his new book, *The Great Art (Ars Magna)*, Jerome gives Tartaglia and Ferrari all credit for their contributions, which, to *his* credit, he lays out with great skill. Scipio Ferreus also receives credit where it is due.

To Tartaglia, however, a credit is not enough. The Stammerer is incensed.

Jerome doesn't understand why, but then there are many things to do with these equations and their solutions that he doesn't understand. The negative solutions, for example, which he calls 'fictitious'. And there are worse cases. There are, for instance, the solutions that involve the square roots of negative numbers — the famed imaginary numbers.

Jerome's *Ars Magna* is the first published acknowledgement that mathematical procedures can produce the square roots of negative numbers. In it, they appear as the solution to a relatively simple problem: 'divide 10 into two parts, one of which multiplied into the other shall produce 40'. The only solutions, Jerome showed, were $(5 + \sqrt{-15})$ and $(5 - \sqrt{-15})$.

Jerome had seen that solving cubic equations often produces square roots of negative numbers along the way. That is because his formula for solving an equation of the form $x^3 = 3px + 2q$ was

$$x = \sqrt[3]{(q + \sqrt{(q^2 - p^3)})} + \sqrt[3]{(q - \sqrt{(q^2 - p^3)})}.$$

Once you start plugging numbers into that formula, square roots of negative numbers can appear very quickly. You can start with very ordinary numbers and sometimes progress through to solutions that contain very ordinary numbers, but you must be prepared to encounter strange and scary beasts along the way. And Jerome was.

Not that he has received the credit he deserved for such boldness. Nowadays, mathematicians tend to credit Descartes with recognising the importance of these 'imaginary' numbers. In his 1637 book *La Géométrie*, Descartes says, 'Neither the true nor the false roots are always real; sometimes they are imaginary.'

By 'false', Descartes means negative numbers, which were themselves considered suspicious and problematic. How much more troublesome the imaginary numbers seemed. Newton called them 'impossible' in his *Universal Arithmetic* of 1707. Newton's arch rival, Gottfried Leibniz, was more positive. In 1702, Leibniz, who was a great admirer of Jerome's work, spoke of the imaginary number as 'a fine and wonderful resource of the human spirit, almost an amphibian between being and not being'.

In the end, it was the Swiss mathematician, Leonhard Euler, who brought imaginary numbers into the mainstream. In the eighteenth century, he followed Descartes' idea and named them imaginary numbers, denoting the square root of -1 as i. He connected i to the real world by showing that a mathematical constant known as e — Euler's number — is connected to pi via i: $e^{i\pi} = -1$.

That e ties together with i and pi is one of the great mysteries of the mathematical universe. e was being used in myriad calculations — from calculations of compound interest, to the power of cannons — and so it wasn't long before Euler's work turned Jerome's 'useless' numbers into an essential component of a mathematician's toolkit. By the end of the eighteenth century, they were required everywhere.

If I want to take something that varies over time, for instance, and calculate its exact value at a certain time, I need the imaginary numbers. That is because they exist in the formula and, as soon as the formula involves squaring a number, they become real. For all their 'imaginary' nature, if you don't put them in, you get the wrong answer.

So, when *are* the imaginary numbers required in a formula? The answer is when there is more than one dimension to a problem — which, in the real world, is always the case. Say I wanted to calculate how fast a team of oxen could plough a field. It is not just about how much power they apply to the plough — there is also the issue of the soil's resistance to their movement. And the amount of resistance changes depending on how fast the plough is moving. It is a complex problem and it requires complex numbers. And 'complex numbers' is the name we give to the combination of real and imaginary numbers.

How does ploughing a field relate to the solutions for cubic equations? Well, I can plot a graph of the speed of the oxen versus the resistance of the soil to the plough. Because the soil's frictional resistance to the ploughshare depends on the speed of the oxen, it wouldn't be a straight line; it would be a curve. And if I wanted to add in a third factor — the change in resistance as the spring sunshine dries out the soil, say — then I would have not just a curve, but a three dimensional curve. That is a solid object, essentially, something like a curvy cone. If I want to know where this curvy cone intersects with another factor — the availability of labourers during the day, say, so that I can work out the most efficient time at which to start ploughing — then I am trying to find solutions to where the curvy cone meets another, related curve. I am, in fact, looking for the solutions to a cubic — possibly quartic — equation. We are back in the territory of Omar Khayyam and Apollonius of Perga.

No one wanting to plough a field in Jerome's day would think about consulting a mathematician. However, people calculating the interest owed on loans certainly did. Bankers and loan sharks lent money with property as security — your house, perhaps, or the contents of a grain-storage barn. With the value of the property changing on a weekly or monthly basis, depending on the economic climate, and accepted interest rates varying on a daily basis, those calculations involved solving what were, at the time, rather complex equations. The bankers didn't need to think in terms of conic sections because the educated men they hired as in-house mathematicians generally had plug-and-play formulae at their disposal. Except, of course, where they didn't. In those cases, a new solution method would, quite literally, put money in the bank.

Things are no different now. When I was doing my PhD I worked with a colleague, Daniel, who eventually pursued a career in finance. His abilities in solving 'differential equations' — the name given to equations that involve a cohort of smoothly varying factors, all of which can change the outcome — have earned him a fortune. His variables are not mud and ploughs, sun and workmen, but commodity prices and shipping times, along with the minutiae of supply and demand. This is why, for decades now, the world's banks have been mopping up some of the best universities' mathematicians and physicists. All of these people know the value of i. To them, it is priceless.

ψ

'And why didn't *you* work for the bankers?' Jerome looks genuinely puzzled. 'You know quantum theory, too. You can solve these equations. You could have made your fortune.'

I raise my eyebrows and smile. 'I could ask you the same thing, surely?'

'I never cared about money,' Jerome says with a shrug.

'Me neither. Otherwise I wouldn't be a writer, would I?'

Jerome acknowledges the joke with a grin, but says nothing.

'And I would never have discovered all of this.' I tap my stone on the *psi*. 'I would have had all the knowledge and none of the wisdom. Daniel can solve equations for money, but he doesn't understand the nature of reality.'

'And you do?' He knows the answer.

'Not yet,' I reply. 'But I'm ever hopeful.'

I can see the smile twitching at the corner of his mouth. 'And is it the Copenhagenists that give you such optimism?'

He is being sardonic, and enjoying it. Jerome had very little time for the Danes. He was once asked by King Christian III of Denmark to travel to Copenhagen and become the royal physician for a salary that was twice what he was earning at Pavia — plus a house, servants, and horses. Jerome politely declined. The Danes, to his mind, were uncultured barbarians and their climate would be the end of him. What, he later wrote, is the use of riches and comfort when the cold and the damp are 'an entrance to death's caverns'?

'I told the King of Denmark that I couldn't help him because I couldn't be in two places at the same time,' Jerome says. He looks thoughtful. 'I said I was a widower with children and that I had to oversee their education. I couldn't be in the north when my heart was in the south.'

'That was an expedient excuse, though, wasn't it? Your children's education didn't stop you going to Scotland, after all.'

Jerome laughs, a low chuckle that makes his shoulders shake for a moment. 'No,' he says. 'But Archbishop Hamilton needed my help more than King Christian.'

'And now you need his.'

Jerome picks up his pen. 'Yes,' he says. 'And now I need his.'

Chapter 8

What to deal with first? That Jerome is a widower — the death of poor Lucia, the much loved wife about whom he wrote so tellingly little? The children who brought their father such misery? Or Archbishop Hamilton?

The Archbishop? At least that story has a neat beginning and a happy ending. Let's start there.

It is 1551. Jerome is entering his sixth decade and is the talk of Europe. He has cured cases of tuberculosis by prescribing clearer air outside of the city. He has healed the tetanus-stricken son of a Bolognese Senator. By 1539, twelve years after their first denial of his suitability to practise medicine, the Milanese College of Physicians had admitted him to their ranks. Two years later, he accepted the offer to become their rector. It is a position of such status in Europe that it brings Jerome to the attention of foreign powers — and, in particular, to the attention of the medical team responsible for Archbishop Hamilton.

Born a bastard like Jerome, Hamilton has risen to become Archbishop of St Andrews and Lord High Treasurer, Comptroller,

Collector General, and Treasurer of the New Augmentation of Scotland. However, he is a martyr to his debilitating asthma. His personal physician, John Cassanate, has abandoned trying to come up with a cure. Instead, Cassanate has used Scotland's status as a solidly independent Catholic state to engage the services of both the doctors at the court of the King of Catholic France and those who attend the French King's nemesis, Holy Roman Emperor Charles V. However, none has managed to be of the slightest help.

Then Cassanate hears great things of one Jerome Cardano. On 28 September, he writes a letter to the Milanese doctor. It takes six weeks to make its way through a violent, war-ravaged Europe. Eventually, though, it is delivered into Jerome's hand.

The letter announces, somewhat presumptively, that the Archbishop will travel from Scotland to Paris to receive treatment from Jerome. All expenses will be covered by the letter's bearer. Flattered, Jerome sets out from Milan on 12 December 1551. He heads due west into France to begin the journey to Paris, but is stopped in his tracks after three hundred miles, in Lyon. By this time, word of his journey has reached Lyon's governor. He is met outside the city walls and told, with flourishes and more flattery, that he simply cannot take another step further — not until the governor has had time to arrange a proper civic reception for the esteemed Doctor Cardano.

Jerome is put in a decorated carriage, with outriders and postilion, and, when suitable preparations have been made, paraded through the city. He is not immune to adulation and he happily stays six full weeks in Lyon, consulting with local dignitaries and offering advice to their physicians. What's more, a second messenger has arrived from the Archbishop and asked Jerome to wait in Lyon for Doctor Cassanate, who will personally accompany him to Paris.

While he waits, the sick of Lyon flock to him, paying good money for his attentions. He even receives a lucrative job offer. If he will remain in the city, a local military commander called Marshal Brissac will hire him as mathematician and designer of artillery. He stands to earn one thousand crowns a year. He turns down the Marshal's generous offer, however, because history beckons.

Jerome is about to step into one of the great intrigues of the century because Archbishop Hamilton is the brother of James Hamilton, Regent of Scotland. James's right to the regency is fiercely disputed, however, with a Cardinal Beatoun vying for the position. To shore up his position, James has made an alliance with King Henry VIII of England: that Mary, the daughter of Scotland's last king, James V, would marry Henry's son, Edward. James will send Mary — to be known as Queen of Scots — to London when she reached the ripe old age of ten. With her would go six hostages who would ensure every part of the pact is upheld.

The Scottish barons hated what they perceived as this weak, subservient attitude to England and conspired with Beatoun to take Mary and her mother, Mary of Guise, captive so that the promise could not be fulfilled. And so the Archbishop encouraged his weakling brother to declare Beatoun an enemy of the state.

Scotland entered what was, effectively, a civil war. Beatoun was murdered, then Henry VIII died in 1547, putting the young (and wifeless) Edward VI onto the English throne. When Edward's Protector, the Earl of Somerset, then invaded Scotland, the Scots asked France for help fighting the English and offered little Mary's hand in marriage to the Dauphin. So it was that Mary, now six years old, travelled to France in 1548 with the returning fleet that had brought six thousand French soldiers to Scotland.

Three years later, in 1551, France negotiated a delicate peace between England and Scotland. However, Mary of Guise, now residing with her daughter in the French court, has designs on the Regency of Scotland. The weakling James Hamilton has clung to the position, but only through the strength and advice of his brother, the Archbishop John Hamilton. And now the Archbishop's asthma is making it all but impossible to continue this role. The fear is that, if Hamilton's health doesn't improve, Scotland will quickly become Mary's dominion.

By the time Jerome is in Lyon, the asthma attacks are coming weekly and lasting for a full day each time. Regent James Hamilton, who feels his brother is not in any state to stand by his side, has promised Mary of Guise that she can have the Regency by January 1552. She is well on her way to achieving all her ends. Only John Hamilton stands in her way.

Which is why the Archbishop cannot afford to leave. Instead of coming to Paris, his personal physician will come to Lyon and escort the good Doctor Cardano all the way back to Edinburgh. It looks like it will be worth Jerome's while. For a cure, Hamilton says, he is willing to pay 'all the riches of my revenues'. Less than twenty years after finding himself homeless and spat upon in the stinking streets of Milan, Jerome looks set to be celebrated with pomp and public acclaim. And lots of gold, of course. The next few months will be the most lucrative of his career.

On the journey north, Jerome is subjected to further fawning. Although he is keen to reach his destination as quickly as possible, the luminaries of Paris will not hear of it. He is given a lavish civic reception and a tour of the city, and persuaded to speak at several specially arranged medical conferences. He is given a chance to wield the sword of St Joan, which is surprisingly heavy — 'an indication of the maiden's unusual strength,' he said. He is also shown the talon of a griffin, which he doubts is real. His most memorable moment, he says, was at the church of St Dionysius, where he was allowed to handle a unicorn's horn. This, he later writes, was truly impressive, 'perfectly smooth … perfectly straight'.

I can't let it go. 'You know it wasn't really a unicorn horn? It's from a narwhal — a kind of whale.'

Jerome looks affronted. 'How do you know?'

'Unicorns aren't real. They're just mythical. Like the griffin. The horn probably came from some sailors making a huge profit from a lucky catch.'

'All of the great minds, writers of some of the most reliable books, tell us tales of the unicorn. What makes you so sure?'

What to tell him? Once, on a guided tour of Vienna's Museum of Fine Arts, I was shown a ruby- and diamond-encrusted goblet. Crafted from horn and gold, it was made for carrying powdered unicorn horn. At the time the goblet was made, the powdered horn was twenty-two times more expensive than gold, weight for weight. But microscopic analysis of the residues inside such vessels has shown them to contain nothing more exotic than narwhal horn.

On the face of it, Jerome is fairly good at sifting evidence and using critical thinking to come to reasonable conclusions. 'It is the role of wisdom to put forward first of all the petty doubts,' he writes in *On Subtlety*, 'and even if this can be done, then to bring up useful solutions and the cause, and to say nothing ludicrous about the presentation of the cause.' He offers useful takedowns of those who claim to have made perpetual motion machines, and even uses his faculties to pre-empt Darwin's 'survival of the fittest' idea concerning evolution in the natural world. His thoughts about the origins of the abilities and physical characteristics of herbivorous beasts include this gem: 'being foolish and timid, they need to be fleet of foot to survive ... those that cannot be speedy do not establish a species'. And yes, in 1550, he uses the Latin word 'species' to talk about a class of animals. According to etymological sources, the word became used in the 1560s to describe a distinct class (of something) based on common characteristics, and its biological use began around 1600. Jerome is an impressive innovator.

And yet, despite this surprising level of insight, Jerome often falls flat on his face — especially where animals are concerned. In 1605, Francis Bacon will publish a book called *The Advancement of Learning* in which he will berate Jerome — and Pliny and the Arabian diviners, to be fair — for lacking rigour in the study of natural history. 'There hath not been that choice and judgement used as ought to have been,' Bacon says. There is, in Jerome's writings, 'much fabulous matter, a great part not only untried, but notoriously untrue ...'

Bacon has a point. Jerome opines, for instance that Lithuanians 'are the most voracious of men' thanks to the influence of

Lithuanian wolverines, creatures with 'the size of a dog, the face of a cat, the back and tail of a fox, rough and tough feet and claws, and teeth too'.

Where the observations appeal, Jerome eagerly swallows and repeats second or third hand reports. Pliny's observation that 'an elephant making its way onto a ship demands an oath that it will return' is one. In fact, Jerome seems to be utterly credulous where elephants are concerned. They are, he tells readers of *On Subtlety*, 'sensitive to pity, and revere a king'. Moreover, elephants 'recognise and enforce an oath, and worship the stars, and feel distress for themselves; they recognise their driver, and extract vengeance from those who have harassed them unfairly, and appear to fall short of humanity only in lacking speech, since many human beings appear more savage than elephants in their practice and their movements'.

The bald truth is that Jerome loves elephants, or at least the idea of them. The ideas we love are always subject to a little less scrutiny. So it is easy to imagine him in Paris, impressed by a unicorn horn, reflecting back to that pouch of powdered alicorn he had given the old wizard in the workhouse. And just think: from abject poverty, to a civic reception in Paris; a few ounces of ground horn must have seemed a small price to pay. Why rob him of that happy delusion, when I have my own mythical creatures? After all, I believe the wave function *psi* is a real, measurable thing. For many of my fellow physicists, this is akin to believing in unicorns.

ψ

As it happens, I undertook that trip to Vienna, where I encountered the goblet for carrying unicorn horn (and some of Benvenuto Cellini's finest works), for the very purpose of understanding *psi*. I had been invited to attend a conference at which some of the leading lights of quantum theory would gather to debate what the wave function is — and, by extension, what quantum theory is, really. If we can understand *psi*, the rest should follow. Except that we failed to reach our destination. How could we not fail, in such a few short days, when the debate was almost one hundred years old? But I had been hopeful and I left Vienna dejected.

In the time since, I have come to think about the wave function like a game of poker. At the card table you make predictions based on probabilities. But those cards are real, physical objects. Just because I can't predict which card will come out next doesn't mean the cards aren't real. I just don't have all the information about their order in the shuffled pack. In the same way, *psi* is real for me.

The fact that *psi* is constructed of real and 'imaginary' parts is not a problem, either. Once you accept that this construction is the mathematical setup that works to explain all the observations, you can get over the fact that the setup is so different from the everyday, 'classical' world.

It's not as if we don't have other areas of expertise where there are real and imaginary parts. One of the reasons quantum theorists are so valuable to city bankers is because many commodity prices are constructed from two (or more) things that work in very different ways — sometimes those things are made up, too. Take silk, for example. A silkworm produces silk in volumes that silk farmers will tell you depends on the ambient temperature.

But temperature is not a physical property of the environment. Temperature is a construct: a shorthand for the kinetic energy available to the atoms in the environment. Temperature tells us nothing about the amount of energy each of those atoms actually has, only what the average amount of energy might be. So the price of silk depends on something real — such as the number of available silkworms — and something abstract — the temperature at which they are housed. I think of these as real and 'imaginary' parts. I could describe the weather and the weather forecast in similar terms. While Niels Bohr might prefer to think that *psi* is nothing more real than a forecast — a declaration of information about a potential state — I consider it to be the weather itself.

Feel free to disagree. Plenty of people do, and I with them, and so we go on, no closer to resolving this argument that Jerome unwittingly started by opening up the worlds of probability and imaginary numbers. In Vienna, I sat at tables where grown men and women would agree on the facts of an experiment and then disagree entirely about what had actually happened. The most belligerent attendee was an Israeli, Lev Vaidman, who believes that the answer to everything lies in a multiplicity of universes: the Many Worlds interpretation of quantum theory.

As a writer, I owe this interpretation a great deal. It was the subject of my first piece in a national newspaper. The Many Worlds interpretation of quantum theory is really quite simple. It says that the wave function exists in that abstract, infinite dimensional Hilbert space that Jerome likened to his *aevum*. And so, if the space has infinite dimensions, there is no shortage of possible outcomes for the processes of the universe. In the double slit experiment, we concern ourselves with the wave function of a photon moving towards a double slit because that is all we are aware of. However, in Hilbert space that wave function is encountering all the other wave functions describing everything else — from atoms in the light-detecting screen, to stray atoms of air in the apparatus, and photons of heat radiation emanating from the material in which the slits have been cut. These intertwine with all the wave functions of all the stuff in the room and beyond, forming something we can talk about as equivalent to the wave function of the whole universe.

The outcome, from our point of view, is an interference pattern that looks as if the photon has gone through both slits. According to the Many Worlds interpretation, however, the outcome is the result of interactions between the wave function of the photon in various dimensions of Hilbert space.

Essentially, all possible quantum eventualities can and do occur — but in the infinite Hilbert space, not in a real (to us) physical space. Many Worlds is often misrepresented as an extravagantly wasteful interpretation where an infinite number of universes sit alongside the one we inhabit. It's neater than that, though. The infinity of universes only exist in an abstract mathematical space. In this worldview, my PhD research didn't involve a current going round a ring of niobium metal in two different directions at once. It only looked that way. In reality, the current was going clockwise in some dimensions of Hilbert space and anticlockwise in others.

The man who came up with this provocative idea was Hugh Everett III, a fiercely intelligent, impetuous, and ultimately self-destructive young man. Born in 1930, Everett grew up to be a genius computer programmer who had almost top-level security clearance in the US military machine. It was Everett who wrote the code for American computer simulations of what would happen in an all-out nuclear war, causing one high-ranking general to comment that Everett was 'worth his weight in Plutonium 239'. It was also Everett who dreamt up Many Worlds. That was in the 1950s and he went to his grave three decades later believing he had found an answer to all the Copenhagenist fantasies. There is no need for a wave function collapse, he said. There is no need to talk about the role of an 'observer'. He could account for the experimental results and the probability distribution that arises from quantum measurements. The only price to pay was a belief in the existence of a near infinite number of abstract universes.

Everett wasn't bothered by the outlandishness of his idea. 'We do not believe that the primary purpose of theoretical physics is to construct "safe" theories,' he wrote in 1956, directly challenging Bohr and the Copenhagenists. The 'collapse' of the wave function's superposition was a 'philosophical monstrosity' to Everett, 'hopelessly incomplete' and 'overcautious'. To Everett, quantum theory could only be explained by a bold idea — and his certainly was bold.

You may not be surprised to learn that Everett convinced few of his peers. So few, in fact, that he went to his grave a broken man, beaten down by depression and addiction. In part, that was down to the stubbornness of Niels Bohr, the great Dane. Everett's mentor was a Princeton physicist called John Archibald Wheeler and Wheeler was incapable of living life without Bohr's approval. So when Bohr refused to countenance Everett's ideas in any way,

Wheeler ended up publicly denouncing Everett, crucifying the younger man in order to keep Bohr onside. The rejection and the sniping drove Everett to a dark place deep within himself. He died of a heart attack aged just fifty-one, a chain-smoking alcoholic who left behind a trail of devastation in his family.

Yet, despite their shaky start, Everett's ideas are now venerated by some of our greatest minds. The idea that the world branches every time we measure the position of an atom, the spin of an electron, or the energy of a light photon sounds outlandish. But why not?

One argument against this theory is that it requires multiple copies of you. After all, there is a universe where an ion moved left to right across a particular synapse within your brain this morning, and there is one where it went in the opposite direction. That moment alone created two versions of you.

In which universe are you conscious? Or are you *you* in all of them? There is simply no way to deal with the consequences of the Many Worlds interpretation without breaking through the limits of what we understand. What's more, it is an untestable idea because we can't access the other branches of this 'super-universe', 'multiverse', or whatever else you choose to call it. It's a fascinating place nonetheless. It is somewhere where everything happens — for anything that can happen will happen because the infinite possibilities are all occurring with some finite probability.

So what should you think about the Many Worlds idea? That is enormously difficult to answer. As with all the other interpretations, it is largely a matter of taste. Can you swallow the idea of multiple copies of you (or at least your wave function) circulating in myriad disconnected, abstract worlds? If you can, you are also signing up to a universe — or multiverse — where a team of monkeys has sat at a typewriter and written the complete works of Shakespeare. And one where you can play Russian roulette and never die.

This latter universe was the subject of my newspaper debut, an article in *The Guardian*. Essentially, it involves a macabre version of the double slit experiment. Instead of a photon being in two places at once, we have a bullet being in two states at once: fired and not fired. And whoever holds the quantum gun to their head is both dead and alive. It was Max Tegmark, a perfectly well-respected quantum physicist, who worked out that this experiment would provide proof that a near infinite array of universes does exist.

Tegmark's quantum gun only fires a bullet fifty per cent of the times that you pull the trigger. The bullet's emergence from the chamber is the result of a quantum action — a measurement on whether a lump of radioactive rock has emitted a particle, for example. If it has emitted the particle, the bullet fires into the subject's brain, ending their consciousness in that branch of the multiverse. But, according to the Many Worlds idea, there is always a universe where the rock hasn't emitted anything. So for every pull on the trigger there is a universe where the subject never dies. You could never prove it to anyone else, but the surviving you would know — if you went through with it — that every event that involves you and a quantum particle will cause your life to play out in a succession of new and different realities. In all the others, your consciousness is shut off. So if the Many Worlds interpretation is right, you have an ongoing experience of immortality. That immortality, Tegmark says, would be your proof of the correctness of the interpretation.

But, he adds, it's just a thought experiment: don't try it. What's more, it is a limited proof. Just as no one else can be sure that you are conscious (because consciousness is a subjective experience), no one else would believe your claims to never once have died while playing a quantum version of Russian roulette.

Tegmark does try to live with the multiverse as a reality. He has said that he feels 'a strong kinship with parallel Maxes', even though he never gets to meet them. 'They share my values, my feelings, my memories — they're closer to me than brothers,' he says. Maybe that's why he hasn't yet performed one hundred rounds of quantum suicide to test the hypothesis.

I suspect that we are living in a universe where David Deutsch, another eminent Many Worlds proponent, has done it, though. How else could he be so sure that these other worlds exist? He once told a colleague of mine that he is as sure of the existence of the multiverse as he is of the past existence of dinosaurs on Earth. This is the man who drew up the blueprint for the quantum computer, a machine that is already changing the face of technology in the twenty-first century.

He's no simple-minded fool. And yet …

If you find Copenhagen too oblique and Many Worlds too fantastical, there are other interpretations that may be more to your taste. We'll get to those. In the meantime, we ought to return to the world where, on 29 June 1552, Jerome finally arrives in Edinburgh.

$$\psi$$

Jerome's journey has been interesting, but largely uneventful. He travelled from Paris to Calais via Rouen, which, he says, has fine architecture and handsome inhabitants. Rouen, to his mind, is a beautiful city: 'only Rome is even better situated and still more imposing and beautiful'. It is, he says, altogether a better place than Paris, which is 'a rather filthy place, filled with stench and poisoned air'. In Dieppe, he marvelled at a gooseberry bush. Once across the channel, he says little of England, but the weather is clearly not to his liking. In *On Subtlety*, Jerome offers a list of twelve celebrity intellectuals who have impressed him. Two are from what he calls 'the part of England known as Scotland' and he remarks that 'the Barbarians are not our inferiors in talent' despite living 'under a foggy sky'. He also speculates on whether the cold of the region stimulates hair growth. 'I remember seeing a younger man in England whose front-body was to my amazement completely covered with hair,' he says. 'In Scotland I saw another man who had hair over his entire body; one would have called him a big bear.'

To further his understanding of the locals, Jerome has purchased *The History of Scotland*, written in Latin by Hector Boece. Within its pages he — and, a few decades later, Shakespeare, via Holinshed — first encounters the macabre tale of the Scottish king, Macbeth. Jerome is also enthralled by the 'most wonderful' fortitude of the Scots, who take a piper with them to their executions: 'he, who is himself often one of the condemned, plays them up dancing to their death'.

For just over two months, Jerome attends to the Archbishop, changing his routines and regimens. He writes copious notes about the causes of the breathing difficulties and the best course of action. He alters the Archbishop's bedclothes, swapping his feather-stuffed leather pillows for seaweed- and silk-filled linen. He supervises the preparation of His Grace's meals and beverages, and oversees his exercise routines. And it seems to work.

Within two weeks, the Archbishop has declared himself delighted with progress and let everyone know that Jerome has put him on the road to recovery. In fact, the doctor's spare time is now taken up with other paying customers, for the Scottish nobles all want consultations over their ailments. Then comes the letter from King Edward VI of England:

It has come to Our ears that you, the great and beloved physician Gerolamo Cardano, have raised from a death bed our holy Archbishop of Scotland and have a great skill in treating these ills of the chest. Therefore at your pleasure wait upon us when you return to London.

Though summoned by a king, Jerome does not leave at once. First, he ensures that the Archbishop and his entourage will be able to continue the various prescriptions in his absence. For the entourage he writes a forty page document and supplements it with a personal handbook for the Archbishop's own eyes. Hamilton, grateful but not quite satisfied, wants more. He has heard about Jerome's skill as an astrologer and asks for a personal horoscope.

Here, Jerome is less happy to help. He has just spent two months putting the Archbishop's body back into action and he knows the power of placebo — or rather nocebo, the negative effect that an emotional upset can have on health. The Archbishop's horoscope might dampen his spirits and a depressed man is more prone to illness. Though he hasn't yet cast it and doesn't know what it might contain, he does not want the horoscope to undo all his good work.

So, when he eventually casts the horoscope, Jerome tells the Archbishop some — but not all — of what he sees in the charts. His Grace, he says, will find happiness — but only through worry and danger. There is no immediate cause for concern, but he might find his life in peril in two years time, in 1554, and again in 1560.

Much like the wiliest of weather forecasts, the prediction couldn't ever be proved wrong, but it is certainly not an accurate depiction of future events as Jerome saw them. That is clear now, in 1570, when Jerome's heart tells him Hamilton is already dead, and cannot help him out of this prison cell.

'You, presumably, know the date of the Archbishop's death?' Jerome is looking at me with a peculiar stare.

'I should, but I don't,' I say. 'I never paid it much attention. 'I just know he is alive now, as we sit here.'

I have looked it up since, of course. As Jerome and I talk, the Archbishop is fifty-eight years old and less than six months away from being hanged for his role in the assassination of Scotland's Regent Moray.

Jerome looks pleased. Not the kind of pleased that results from the realisation that his life might be spared; more the kind of pleased that he did a very good job of being the Archbishop's doctor. Smug, you might say.

'He was a very sick man. That is one of my greatest achievements,' he says.

Smug, but correct. Indeed, the Archbishop's condition was so improved that he begged Jerome to stay until April, promising to lavish him with even more gold than the doctor had already amassed. Jerome, flattered but genuinely unmoved by promises of gargantuan wealth, replied that he was too homesick to stay any longer. And so — with horses, chests of gold, and a cavalry escort — he set off for London.

'You liked the English, didn't you?'

I have read his description of the sixteenth-century inhabitants of London, published posthumously in his *Somniorum Synesiorum*. They seem to him 'urbane and friendly to the stranger', like white-skinned Italians. His observation is that, as well as being hairy, they are broad chested, quickly angered ('and in that state to be dreaded'), almost as gluttonous as the Germans and utterly unafraid of death. 'With kisses and salutations parents and children part,' he writes. 'The dying say that they depart into immortal life, that they shall there await those left behind; and each exhorts the other to retain him in his memory. Cheerfully, without blenching, without tottering, they bear with constant the final doom.'

He finds their language impenetrable. 'When they opened their mouths I could not understand so much as a word,' he admits. 'For they inflect the tongue upon the palate, twist words in the mouth, and maintain a sort of gnashing with the teeth.' It leaves me wondering how, then, he gained any impression of the English. But then I am also wondering how it is that he and I are able to converse. Perhaps some things are beyond mere words?

ψ

If Scotland is a political mess, London is no better. It doesn't take long for Jerome to realise he has walked into a spider's web: almost immediately he is offered a thousand gold crowns to treat the young King Edward and endorse his new title as Defender of the Faith. Jerome extricates himself from this sticky situation with unusual tact and diplomacy. He is a subject of Italy and thus of the pope, he says: his endorsement counts for nothing. Such integrity costs him dearly, for the fee on offer drops immediately to one hundred crowns. He is invited to the French ambassador's house, where he is offered a stipend of eight hundred crowns a year if he will walk away from the royal court in London without treating the king. Jerome politely declines, as he does with emissaries of the Emperor Charles V, who also wants the celebrated Doctor Cardano to spurn the English heretics.

The manoeuvring doesn't stop there. He is only allowed to examine the king after a few weeks of hanging around. There is nothing terribly wrong with the boy, is Jerome's diagnosis. That is when he realises he has been brought here for other purposes. Pressed for the king's horoscope, Jerome decides that the pragmatic response is to obfuscate.

Edward VI had been crowned King of England at the age

of nine on the death of his father, Henry VIII. He is now fifteen years old. Although Edward is not particularly unhealthy during the time of Jerome's visit, measles and smallpox have made him prone to illness and the courtiers want to know a date of death and a foretelling of the political consequences.

Here is the dilemma: it is clear to Jerome that the royal court is preparing for chaos. With the king too young to rule alone, England is already governed by a Regency Council that is led by noblemen loyal to Henry Tudor's Protestantism. When Edward dies — and most believe his demise is imminent — the attempt to keep the pope out might be undone by his Catholic half-sister, Mary, who could reasonably ascend to the throne. Her claim would be supported by France and Spain, and her ascension would be a severe blow to the Reformation in England. If he says the wrong thing, Jerome could find his exit blocked. If he says the words English ears want to hear, however, he could find himself persona non grata back in Catholic Italy.

So Jerome procrastinates. He spends time with Edward, whom he clearly liked and admired. 'He was a marvellous boy,' he wrote later. 'I was told that he had already mastered seven languages. In his own language, French and Latin, he was perfect.' Jerome noted that Edward was quite open to being taught and quite capable of holding a philosophical discussion in Latin — and of asking insightful questions.

Not being able to escape the situation forever, Jerome eventually draws up Edward's horoscope. He gives a plausible but generally positive account:

At the age of twenty-three years, nine months, and twenty-two days, languor of mind and body would afflict him.
At the age of thirty-four years, five months, and twenty days, he would suffer from skin disease and a slight fever. After the age of fifty-five years, three months, and seventeen days, various diseases would fall to his lot. As long as he lived he would be constant, rigid, severe, continent, intelligent, a guardian of the right, patient in labour, a rememberer of wrongs and benefits; he would be terrible, and have desires and vices growing from desire, and he would suffer under impotence. He would be most wise, and for that reason the admired of nations; most prudent, magnanimous, fortunate, and, as it were, another Solomon.

And then he runs off home. Within a year, King Edward VI is dead and England is in turmoil.

'Did you see he was going to die, and thought it best not to mention it?'

'Are you telling me you believe in the powers of astrologers?' Jerome allows himself a sly grin. 'It's hard to know what I saw, really,' he shrugs. 'But there was nothing to be gained by predicting his imminent demise. We astrologers have a long tradition of not making such pronouncements. There are uncertainties in our information about the future and sometimes we feel we shouldn't trust ourselves to convey it well.' He pauses and points a finger at my scrawls. 'Are you certain of everything you tell me about *psi*? Would you risk someone's life on your interpretation of its powers? Would you risk your own life in order to vindicate the wave function?'

I don't answer. I'm certain of nothing to do with *psi*. This wave function may not represent anything beyond a mathematical tool for making bets on quantum measurements. There is no certainty to be had here yet. And as for the possibility of information from the future being known in the present, well, I know a very eminent Israeli man who believes that is possible. His name is Yakir Aharonov, and he has a National Medal of Science that was bestowed upon him by the President of the United States, Aharonov's adopted home. It's almost funny: Jerome worried, on hearing about Christopher Columbus's journey, that the discovery of the Americas was a harbinger of the end of time itself. It is a moment that is 'sure to give rise to great and calamitous events,' he writes in his autobiography. Inadvertently, he may have been right.

$$\psi$$

Aharonov's interpretation of quantum mechanics is as much of a mind-bender as any idea about time travelling must be. But, he argues, if it's a mystery to have a photon be in two places at once, it's perhaps even more of a mystery to have the photon know whether it's being watched.

The issue here is that experiments have shown that if we put a detector on one of the slits in the double slit experiment, the interference pattern disappears. Why? Because, before it even gets to the slits, the photon has seen that it can't pass through unnoticed.

Aharonov solves this with a simple tweak to our understanding. The photon's wave function, he says, is put together by information from the future as well as from the past. In the future, the photon will encounter the detector on the slit and this changes the wave function so that the superposition collapses before it has even formed.

You might, at first glance, find this a bit hard to swallow. Aharonov, though, says this 'solves the measurement problem'. And surely that's a reason to take it seriously.

Aharonov is one of the physicists whom I met in Vienna. He is a giant of physics, quite simply one of the most brilliant people, one of the clearest thinkers, of our day. It may not always be easy to understand what he's saying — he has a speech impediment that makes me think of Nicolo Tartaglia — but when you get it, it's *very* convincing.

So here's what Aharonov says: forget what you think you know about time. For starters, it's not about a progression of moments. Instead, time is actually about the flow of information. For us, it flows only forwards: we have information from the past, but we can't access information from the future. However, the deeper reality is that time — and, by that, I mean the flow of information — runs both forwards and backwards. And if you know how to access that future information, you can resolve the mysteries of the present.

You're probably thinking of astrology. Aharonov is thinking of two controversial ideas in quantum physics called weak measurement and postselection.

A weak measurement is a measurement that doesn't disturb

the system enough to change it significantly. Imagine you want to learn the mass of a bicycle that is moving at speed. To measure it properly, you'd have to stop it and put it on a set of scales. But you can gain a tiny sense of its mass by giving it a little sideways nudge as it passes. It will deviate from its course for a moment, but it recovers and continues forward. The amount of deviation, compared with the effort you put into the nudge, will have given you some sense of the bicycle's mass. That's a weak measurement. The point about a weak measurement is that you gain a small amount of information that is clothed with a huge amount of uncertainty. In other words, you have learned something, but you're not sure whether it is true. However, if you probe the system enough times, you may achieve a consistency to the answers that convinces you of their essential correctness.

Applied to measurements on a photon in a double slit experiment, Aharonov says, that gives you a new capability. Throughout those weak measurements, the photon's state is a mix of its past state and its future state. In other words, it contains information from the future. And so your weak measurement is recording a tiny sliver of what the future holds.

But for Aharonov, that is not quite enough heresy. He also advocates discarding experimental results that you don't want.

'You can't do that,' Jerome says. His eyebrows have arched, and his jaw drops. 'It undermines the whole point of doing an experiment.'

Perhaps you are shocked that Jerome is shocked? Somehow in the last couple of centuries we have convinced ourselves that experimental science only began in the century after Jerome died. We think of Newton performing experiments that show the dispersal of white light by a prism, or of Galileo dropping cannonballs from the tower at Pisa to test the nature of gravity (he almost certainly didn't, but he did perform other related experiments).

What we haven't absorbed, largely because of cultural bias, is the long Islamic history of experimentation. In 1021, for example, the Islamic scholar Ibn Al-Haytham used a camera obscura to prove that light travels in a straight line to the eyes. In the following century, Al-Khāzini's *The Book of the Balance of Wisdom* describes a huge variety of experiments that prove theories of mechanics. A few centuries later, scholars began to translate Islamic scientific and mathematical texts from Arabic into Latin. Inspired, Jerome and many of his contemporaries began conducting their own experiments to test the validity of ideas about how the world worked.

The main difficulty they faced was the need for craftsmen. Every piece of equipment had to be designed and built, sometimes to exacting specifications. That was time consuming and expensive, so experimentation was not for the faint hearted, or the poor. Nonetheless, we possess records outlining the experiments of Guidobaldo del Monte (who helped Galileo acquire his first academic position), the Neapolitan Vicenzo Pinello (another friend of Galileo's), and a certain Nicolo Tartaglia.

These experiments were mostly to do with the balance. Think of the old grocer's scales, where the weight of a measure of flour in one pan is balanced against a set of known weights in the other pan. This was the kind of apparatus by which Tartaglia and his peers worked out the rules governing forces associated with weights and levers. Tartaglia's experimental contributions were nothing special, but his writings do him credit when it comes to the philosophy of science. Tartaglia openly disputed the work of Aristotle, which made no distinction between theoretical arguments about the operation of a balance and the performance of a real world balance. Aristotle dealt with abstract mathematics; Tartaglia points out that when translated into the real world, the mathematics will be subservient to the limits of apparatus and experiment. It might seem obvious to us now, but it was a valuable contribution. 'The physicist considers, judges, and determines things according to the senses and material appearances, while the mathematician considers and determines them not according to the senses, but according to reason, all matter being abstracted,' he says in his 1546 book *Diverse Questions and Inventions*.

For that reason, Tartaglia argued, you have to take into account the weaknesses of your apparatus:

Smaller balances are found to be more sensitive than larger ones. That this is true in material balances, experience makes manifest; for if we have a worn ducat and want to see by how many grains it is too light, using a large balance such as one of those used to weigh spices, sugar, ginger, cinnamon and such materials, we shall get a poor result.

This is, he says:

[Because of the] difference between the material parts
or members of which they are composed, which parts or
members are the two arms and the pivot ... For the said
arms and pivot in the larger scale or balance are much
more gross and bulky than in the smaller. And since
the arms of those scales or balances are to be considered
mathematically, that is, apart from all material, they are
considered and assumed to be as simple lines, without
breadth or thickness; and the pivot or axis is assumed to be
a simple indivisible point.

It is a good point and well made.

Tartaglia, as we know from his book on artillery, also performed experiments related to the trajectories of various armaments. It is more laudable than Jerome's insight into artillery, which was based on observation rather than experiment. Jerome noted that a projectile's trajectory 'imitates the form of a parabola', but he didn't actually investigate it.

Jerome's experimental interests lay in more abstract areas. He made an effort, for instance, to measure the relative density of air and water by shooting a bullet from the same source through the two media. His hypothesis was that the distance travelled by the bullet is inversely proportional to the density of the medium. Water, he concluded, is fifty times denser than air. That's an enormous underestimate and we definitely cannot celebrate Jerome as a brilliant scientist. Sometimes, he even led himself to a false understanding of the world despite carrying out experiments.

Take his encounter with Turin physician Laurentius Guascus Cherascius.

In a gathering with Jerome and some of his friends, Cherascius claimed that touching a needle to a lodestone, a piece of magnetite, would imbue it with anaesthetic properties. 'As one might expect, since we thought this absurd, he confirmed the state of affairs by experiment on my companions,' Jerome writes in *On Subtlety*. Jerome's friends were convinced, so he decided to test this 'incredible state of affairs' on himself. 'I rubbed a needle first on the lodestone and introduced it into the skin of my upper arm. At first I felt the slight impression of a prick; later, since it was making its way virtually straight through the whole muscle, I could feel the needle penetrating into the depths on its journey, but I felt no pain whatever — and then I believed my friends because I had tried it out on myself.'

Don't try this at home. It doesn't work.

'I felt no pain.' Jerome is adamant. His eyes are wide and willing me to accept the truth.

'I believe you,' I say. I do — he wouldn't lie. 'But that doesn't mean the lodestone works. When your mind becomes convinced you will feel no pain, your body releases its natural painkillers. We call it the placebo effect. I expect Cherascius's gift was in talking you into a conviction the experiment had worked.'

Although Jerome looks unconvinced, he says nothing. He has the habit of not challenging certain things that I offer up. Behind his eyes, something tells me he is squirrelling my observations away for contemplation — and use — at a later date. He is a voracious consumer of knowledge. I sense that I am only whetting his appetite.

And to be fair, he is still working out what constitutes knowledge. He is aware, for instance, that experiments can be misleading and that reports can be fraudulent. Experiments, he contends, are best observed directly — a caution that the Royal Society of London would put into practice a century later. He is particularly sceptical about reports of alchemical success. 'A Venetian apothecary by the name of Tarvinius allegedly changed mercury into gold, in the presence of administrative authorities and scholars, and this wondrous occurrence is still remembered,' he writes in *On Subtlety*. 'But whichever way this may have come about, it is quite certain that mercury cannot be changed into gold.'

He also doesn't fall for appeals to authority. He questions Aristotle's claim that air is naturally hot. Discussing the atmosphere in *On Subtlety*, he points out that 'the air in the upper part, where it is not warmed by the reflected rays of the Sun, is cold'. This, he says, 'must cast great doubt on the followers of Aristotle'. In a discussion of a piece of Hippocrates' wisdom, he declares that 'I believe Hippocrates not because it is Hippocrates speaking, but because I am forced by his reasoning to concur in what he says.'

Jerome also balks at 'received wisdom', such as comets heralding great events. He compares the occurrence of significant celestial sights against outbreaks of plague, for instance. Popular wisdom, even among scholars, says that these events are linked, but Jerome sees the truth: 'if nothing else ensues, the sign is misleading', he says. 'In fact in 1531 and the subsequent years there were many comets, and from 1539 right to 1551 so many solar eclipses and great lunar eclipses were seen that more and greater ones are never recorded — yet Italy from 1524 to the present year 1559 has suffered no notable disastrous plague, indeed not even a trace of one.'

But he is a great believer in guardian angels, let's remember. And his reasons for assuming that the Venetian demonstration of alchemy was a fraud have a great deal to do with his belief that stones and metal are animate, organic substances with properties akin to living things. But I don't want to be too hard on him. That belief — and all his others — were always subject to review if the right evidence came in. He built his knowledge through 'inferences from many facts well-known'. One of his self-imposed rules of conduct was to 'observe all things, and not think that anything happened fortuitously in nature'. This is not a path to riches, he ruefully acknowledges (and all working scientists would concur). 'I am richer in the knowledge of Nature's secrets than I am in money,' as he writes.

In short, he knows the dangers of ignoring or discarding inconvenient evidence.

'This Aharonov,' he says, stroking his beard. 'What does he look like?'

It is my turn to raise an eyebrow. 'Why?'

'Because a man's character is written in his face and on his body.' Jerome looks at me as if I am a dullard. 'That is,' he says, 'why humpbacks and squint-eyed men are not to be trusted.'

My eyes widen. 'What?'

'It's true. People say such deformities are caused by witchcraft, but that's wrong: it is simply a consequence of nature's errors. Nonetheless, a humpback has a deformity around his heart and a squint is an error near the brain. Both will affect character.'

I open my mouth to respond, then let this pass. 'Aharonov looks a bit like you, actually.'

Jerome's eyes narrow. 'Then perhaps he is not too dishonest,' he says. 'But I still don't believe in throwing away our hard-earned facts.'

To be fair, neither does Aharonov — not exactly. What he believes in is a method involving what he calls *preselection* and *postselection*. This, he says, is how we can read information from the future.

ψ

Imagine a stream of particles moving around my ring of niobium metal. I want to know each particle's momentum, but I can't measure it directly, as that would affect the result. So I first 'preselect' a particular state for the particles. This is like applying a filter. I'm going to choose to deal only with particles that are moving clockwise around the ring. Next, I perform a series of weak measurements on each of the particles, gently probing the property I am interested in — the momentum — without disturbing them. The final act is to make a strong measurement on each of the particles. Then I look at the state each one is now in. Is it still travelling clockwise or has it been so disturbed by the final measurement that it is now travelling anticlockwise?

153

Now for the analysis to find the momentum of these quantum particles. The first thing to do is to discard all the particles that weren't properly disturbed in the final measurement. I 'postselect', keeping only those that are now travelling anticlockwise. Aharonov showed that quantum statistics can accurately infer their momentum at a time before they were disturbed. And that, he says, contains information about their future state. So postselection and weak measurement are crystal balls that allow us to glimpse the future of our quantum objects, but with the added twist that those future states are real *now*. However, for me, this way of interpreting superposition and the measurement problem feels like a step too far. Am I really meant to get my head around this?

Despite my admiration for Aharonov, and the fact that some of the smartest people on the planet think it's plausible and indeed correct, I find myself unconvinced. I can justify that: some of the other smartest people on the planet think it's a con.

The one reason I think there might be something to Aharonov's idea is that we already know that our notion of time and space is fundamentally flawed. Perhaps it's time to visit the ever-astonishing topic of quantum entanglement.

Chapter 9

The idea that someone or something controls all events goes a long way back. Democritus, for instance, said two and a half millennia ago that any apparent randomness in the world is just down to a lack of information. Faced with quantum theory, Einstein felt the same: 'Quantum mechanics is very impressive. But an inner voice tells me that it is not yet the real thing. The theory produces a good deal, but hardly brings us closer to the secret of the *Old One*. I am at all events convinced that *He* does not play dice.'

Einstein's problem starts with the Schrödinger equation, the one I have scratched onto the wall of Jerome's cell. This describes how the properties of a particle — an electron, say — will evolve over time. Let's consider the position of our electron. The quantum description of that position — given by the Schrödinger equation — cites an 'amplitude', formed from *psi*, which is composed of two components: a set of real positive or negative numbers, and a set of Jerome's imaginary numbers. The particular numbers — real and imaginary — that are relevant to a real world experimental situation depend on the kind of measurement being considered.

All you can do with this amplitude is to specify a position and work out the probability that a single measurement on the electron will give that particular position as the result. When you actually perform the measurement, you might not get that result. But if you do the measurement many times, resetting the system each time, the distribution of results will match the probability-based predictions of the Schrödinger equation. That's why, ultimately, our world becomes deterministic — the predictable pattern is the net result of many, many unpredictable individual events and processes.

Here is Einstein's dilemma. Probability and statistics has produced a useful theory that matches experimental observations. But the theory does not *explain* anything. It can be used to predict what will be observed in experiments, but says nothing about why electrons, photons, and atoms behave as they do. Electrons jump between energy levels, producing radiation at wavelengths and intensities that the theory predicts. But *why* do they jump when they do?

An analogy might be the ancient saying, 'Red sky at night, shepherd's delight.' Humans have known for millennia that an orange-scarlet tint to the evening sky is a herald of sunny weather when morning comes. It is useful and mostly accurate. But it doesn't have anything to say about why the two phenomena are linked. The fact is that a red evening sky does not *cause* a sunny morning. When the saying was invented they were two facets of some mysterious meteorological phenomenon.

Not that it has remained mysterious: we know now that the red evening sky is the result of a region of high atmospheric pressure moving in from the west. This pressure traps dust in the atmosphere and the dust scatters blue light away from our line of vision, leaving only a red glow. Because of the way the Earth turns, dragging the atmosphere with it, that high pressure ridge moves in from the west overnight to be overhead in the morning, and high pressure generally means good weather.

In stark contrast, we have failed to find any mechanism behind quantum events. Einstein thought there must surely be some way to predict when an electron will make its jump and send out a flicker of light. Until we have that in our hands, he reasoned, we cannot call quantum theory 'the real thing'. At the moment of writing that passage to Max Born, in December 1926, Einstein felt quantum mechanics was incomplete.

Now, almost a century later, it is still incomplete by Einstein's standards. Our fundamental belief now is that quantum mechanics operates at random. The Old One does play dice, if you will, which means probability is the only way to quantify what will happen in our experiments. Jerome wrote that his book *On Subtlety* — published in 1550, when he was forty-nine years old — aimed to be his 'complete account of the universe in a single volume'. But it turns out that his *Book on Games of Chance*, written three decades earlier, might have been a more succinct account of the operating principles of the universe.

No one worked harder than Einstein to think through the issue of fundamental randomness. He worried at it for years, formulating thought experiments and seeking acceptable, sensible solutions to the conundrums it threw up. In the end, he went all the way back to the Schrödinger equation and worked with

Schrödinger to try to make sense of it all.

Together, Einstein and Schrödinger alighted on two thought experiments that have become the touchstones for quantum strangeness. They are known as 'Schrödinger's cat', and 'EPR' (where the P and the R refer to Boris Podolsky and Nathan Rosen, the colleagues who helped Einstein formulate his ideas).

Schrödinger's cat goes back to the equation that bears his name. Apply it to a radioactive source, such as a lump of radium, and it gives a probability that an atom within the radium will decay, emitting a burst of radiation, within a certain time. According to the Copenhagen interpretation, until that decay is measured the atom is in a superposition of decayed and not-decayed. In 1935, Schrödinger published a paper containing a novel thought experiment that Einstein considered the 'prettiest way' to show the incompleteness of quantum theory. In it, the radium is placed next to a radiation detector that causes a hammer to fall when it detects a burst of radiation. If the hammer falls, it smashes a vial of cyanide. Next to the vial is a live cat. Should the vial of cyanide be smashed, the cat will inhale the toxic gas and die. All of this is held within a closed box so that no one can tell whether the cat is dead or alive.

Now, instead of a superposition of different photon trajectories, as in the double slit experiment, we have a superposition of live and dead cat. The radium atom's decayed-and-not-decayed state — a reasonable scenario within the normal parameters of quantum theory — has been strung out into an absurdity. Clearly, Schrödinger said (and Einstein agreed), this shows that something is missing from quantum theory.

If it was missing then, it still is now. We could have endless discussions about the question of measurement. Does the cat seeing the hammer fall count as a measurement? If Schrödinger opens the box with his eyes closed, is that a measurement? The simple truth is that no one has resolved this paradox. We simply live with it.

The second thought experiment — EPR — has been resolved, but not in a way that makes the quantum world an easier place to understand. Einstein's EPR paper was published in 1935, the same year as the Schrödinger's cat paradox. Its title was 'Can quantum-mechanical description of physical reality be considered complete?' The answer that it suggested, using a simple setup, is a resounding 'no'.

EPR is, essentially, this: imagine two quantum particles, A and B, that briefly interact, then move away from each other. The way the mathematics of the Schrödinger equation work out — including a significant contribution from the imaginary numbers — means that an interaction between two particles leaves them as somewhat altered entities. They are now one system and the properties of one cannot be separated from the properties of the other.

According to the Schrödinger equation, the interaction creates a new quantum state, one that Schrödinger called 'entangled'. Entanglement, he said, is the 'defining trait' of quantum theory, setting it apart from any other field of physics. Once the particles are entangled, they have no individual existence. Though they might become physically separated, the information needed to describe their various properties fully — and thus to make predictions about the outcomes of measurements on either one — is shared between them.

As Einstein and Schrödinger noted, this had very odd repercussions. Perform a measurement on one, turning its inherent, random potential into a defined property, and you have also affected the outcome of a subsequent measurement on the other. The mathematics of the link between them — the entanglement — means their properties on measurement will be 'correlated'. That might not seem strange at first glance, but delve into the implications and you'll soon see why Einstein called it 'spooky'.

Imagine those two measurements were carried out within a split second of each other, with the particles extremely widely separated in space. So widely, in fact, that there is no way that a signal travelling from one to the other, limited as it would be by the speed of light, can allow one measurement to influence the outcome of the second.

Now let's focus on the details of the measurement. EPR uses the fact that position and momentum share what is known as 'an uncertainty relation'. We'll get into the details of this oddity later; for now, let's simply use the fact that, if we measure the position of particle A precisely, there is a concomitant fuzziness to its momentum. If we measure A's momentum, the position becomes uncertain. Now apply that to entangled particles and the Schrödinger equation tells you that the choice of measurement — position or momentum — on particle A will immediately affect the outcome of a measurement on B. The two are 'correlated', in other words. Instantaneously, B somehow knows which property — its position or momentum — must be precisely defined. That is true regardless of whether A is on Earth and B is on the surface of Mars. A signal travelling at the speed of light — the maximum speed of the universe — would take thirteen minutes to pass between A and B, so there is no way for B to 'know' how it should behave. That's why Einstein dismissed this whole thing as impossible 'spooky action at a distance'.

In their paper, Einstein, Podolsky, and Rosen made it clear how they felt. 'No reasonable definition of reality could be expected to permit this,' they said. They argued that there must be something missing from the Schrödinger equation — that quantum theory was, as yet, incomplete, with some missing information they called 'hidden variables' waiting to be included.

They were wrong, as it turns out.

In 1964, an Irishman called John Bell formalised Einstein's objections and laid out an explicit test for the existence of the hidden variables. He started with pairs of entangled particles — two photons, say, that have interacted at some point in the past. These photons now have shared properties — that is, a full description of photon A involves some of the characteristics of photon B, and vice versa.

According to standard quantum theory, a measurement on photon A produces a random result, something like a coin flip. But according to the Schrödinger equation, the entanglement means a measurement on photon A will instantaneously affect what you'll get as the result of a measurement on photon B — even if they are millions of light years apart. So the result of measurement B is *not quite random* when compared against the value yielded by a measurement on A. Is that because there are as yet undiscovered properties — hidden variables — in A and B that correlate the outcomes of the measurements? Or is there really something in quantum theory that spookily defies our cherished notions of space and time?

Bell thought up a way to tell if the results are determined by hidden variables or by random processes. His setup is complex and subtle, but you can imagine it as a hypothetical team sport, based on choosing a type of measurement to perform and guessing the outcome.

The teams play in two different kinds of universe. Team Einstein makes the choice of measurements in a common sense universe, where there are hidden variables that carry information. These variables are constrained by the laws of physics, meaning their influence cannot travel faster than the speed of light. Here, any apparent magical correlation between the outcome of a pair of distant measurements is an illusion created by the hidden variables.

Team Bohr, meanwhile, is working in a universe where the result of a measurement really is utterly random, but correlations exist between the two particles' properties that will be manifest once one particle's properties become definite.

The setup of the game runs something like this (I'm only going to be able to convey a rough impression — the real world version is fiendishly complicated). Each team is composed of two players, one of whom is on Earth and the other on Mars. It takes thirteen minutes for a signal limited by the speed of light to travel between them. Each of the players has one of an entangled pair of photons. Playing the game involves making measurements: first on the Earthbound photon, then on the one on Mars. To avoid accusations of cheating, the second measurement must happen within thirteen minutes of the first, so that there can have been no communication between the two players, who we'll call Alice and Bob.

Having seen the outcome of the first measurement, Alice then chooses to hold up her left or right hand. Bob, who doesn't know anything about the outcome of Alice's measurement, or which hand she has raised, then performs his measurement and chooses a hand to raise depending on the outcome. There are two ways the pair can score a point: either by raising the same hand when the two measurement results are different, or by raising different hands when the two measurement results are the same.

If you slave your way through all the possibilities, using Jerome's mathematics of probability, you can work out that the best that Team Einstein can do in their common sense universe is score a point on seventy-five per cent of its plays. But Team Bohr, which is operating in a universe with entanglement, can score points on eighty-four per cent of its plays. Why? Because entanglement slightly alters the probabilities. The instantaneous, appear-from-nowhere correlations of entanglement skew the properties of the second photon in a way that makes it slightly more likely that Bob's guess about which hand to raise will be correct. In other words, entanglement means that Team Bohr should always win.

To determine which kind of universe we live in, you just have to play the game. It's possible to do this without going to Mars; you just have to situate your two experimenters far enough apart that the time elapsing between their measurements is less than that needed for communication between them at the speed of light.

The first people to do it properly were Stuart Freedman and John Francis Clauser. They played Bell's game in 1972 and their results showed clearly that we live in Team Bohr's universe. The players scored points eighty-four per cent of the time and there is no way they could do that if they were living in Team Einstein's universe.

Freedman and Clauser's experiment showed entanglement is real. The outcome of the second measurement was influenced by the choice made in the first measurement. The first photons' properties materialised at random when measured. But the measurement choice for the first photon instantaneously affected what would happen when its entangled twin was measured: its properties were not quite as random.

In the years since that first experiment, we have got better and better at exploring the nature of quantum entanglement. A notable landmark came in 2008, when a Swiss physicist called Nicolas Gisin separated his photons by eighteen kilometres. One was in Jussy, a town east of Geneva. The other was in Satigny, a town to the west. The measurements allowed the players to score points eighty-four per cent of the time, and the time between measurements was so short that any influence would have had to travel at ten thousand times the speed of light. Entanglement, that phenomenon that is written into the Schrödinger equation and results from the phase and its imaginary numbers, is indubitably real.

No one knows how it works. The entangled particles are chained together by a connection that we don't understand. They may be one particle that manifests in our world in two separate places. They may even be, by some hidden, contorted geometry of space, right next to each other.

We just don't have any idea of how the measurements can be so well correlated.

ψ

'Fate,' Jerome says.

I shrug. 'Maybe.'

As a trained physicist, you might think I shouldn't believe in fate, but you would be wrong. I can believe that everything is controlled by an external influence, that all things are ordained, and there is no such thing as a choice. Why? Because Gerard 't Hooft can believe it. He has a Nobel Prize in Physics, is universally respected as one of the smartest physicists on the planet, and believes in something called superdeterminism, another selection among the interpretations of quantum theory.

Remember how Jerome once said that wisdom is to 'say nothing ludicrous' about the cause of an effect? I am going to have to ignore that advice. We'll start from the premise that quantum mechanics is nothing more than a mathematical tool that gives you a way to calculate the outcomes of experiments. By the end, you will have lost all free will and will understand that you are nothing but a biological tool of the universe ... Ready?

It begins at the beginning of the universe. There was a Big Bang moment and some of the energy of that moment condensed into matter, taking the form of particles that went on to create atoms. That means the fundamental particles — the electrons, quarks, neutrinos, and so on — all had a common origin. In that case, could their common origin have a lasting effect? What if certain aspects of their properties are, and always will be, correlated?

The consequence would be, essentially, that we are fooling ourselves when we think we are conducting experiments on uncorrelated, independent, entirely separated systems. That is what the superdeterminism interpretation of quantum mechanics says: all those conclusions we have drawn about spookiness or weird superpositions are a consequence of our blindness to the threads that connect everything in the universe.

Superdeterminism doesn't permit randomness. There is always cause and effect. So when a radioactive atom decays, that doesn't happen at random. There is a reason for it — and that reason is tied up in the hidden threads that we can't access.

Similarly, if we prepare two identical atoms in the same state, put them in exactly the same environment, and watch what happens, we may well see two different end results. The Copenhagenists put this down to inherent randomness in the evolution of quantum systems. Superdeterminists say it's because you were fooling yourself when you assumed you had prepared the atoms in the same state. You couldn't have because you have no way of controlling the hidden threads. And the difference between the hidden threads controlling the properties of the two atoms is what gave you two different outcomes. There is no randomness, only ignorance.

Essentially, the superdeterminists' view is that we are not in control of our experiments.

There is no way to separate the settings of a detector, say, from the state of the particles it is about to detect. So you can try all you like to control your test for eighty-four per cent correlations. You can set up and operate your double slit experiment, if you so desire. But who is to say that the atoms of the material that is emitting the photons are not linked to the apparatus used to detect the photons?

Maybe a tweak to the emitter tweaks the detector in some hidden way, giving you the illusory result of an interference pattern at the detector — which makes you think that the photon went through both slits simultaneously, or that entanglement is real.

There is no superposition in the superdeterminists' view. The photon in the double slit experiment is not in two places at once. There is no spooky action at a distance, either. Those are just shorthand descriptions that make sense of the mathematics of quantum theory — and quantum theory, as it currently stands, is not the final answer.

The big problem with superdeterminism is philosophical. It requires that science is an act of self-delusion because all the atoms of the universe are linked in a way that destroys human free will. You're not free to choose how to tweak the emitter because the atoms in the neurons of your brain are also subject to those hidden threads. And, by extrapolation, scientists are not freely choosing the tasks they perform. We are just a part of the complex clockwork mechanism of subatomic physics. Think of yourself as a cog, with teeth that are enmeshed with those of another cog. If that cog turns, can you choose not to?

This idea has been labelled the 'ultimate conspiracy theory'. One researcher once put it to me like this: 'According to superdeterminism, you're allowed to say about any experimental result: "Well, maybe that happened because of a giant, universe-wide conspiracy involving both the particles you measured and the atoms of your own brain — which allowed the particles to know in advance which experiment you were going to do and to get into just the right state, thereby fooling you into thinking that, had you chosen to do a different experiment (which is actually impossible, since you lack free will), you would've continued to see results consistent with standard physical theory. So it all looks like the standard physical theory is valid, but really it's not.'"

It's safe to say, he wasn't a fan.

Gerard 't Hooft is unconcerned. His take on superdeterminism is that you can believe in the other interpretations if you like, but you won't ever have a satisfactory explanation for what is going on. There'll be inexplicable measurement-induced collapses, things in two places at once, and spooky action at a distance. With superdeterminism, you simply assume there's more to the world than we can currently see — there is a God's eye view — and that all the weirdnesses would have a perfectly good explanation if only our experiments were able to probe things down to the ultimate, fundamental scales of reality.

But even then we'd need to use our own brains and they are part of the conspiracy. The brain is a machine made of atoms that have particular properties, and those properties determine what our conscious state is at any moment. Those atoms weren't always inside our skulls: they have been in stars and travelling through intergalactic space. They were part of a water molecule, perhaps, or once skirted the event horizon of a black hole on the other side of the universe. Ultimately, they can trace their origins to the moment after the Big Bang, when the universe first saw matter. And the imprints that one atom left on another back then continue to exert their influence now, even as carbon-based life forms attempt to construct stories about how the universe really works. Yet even 't Hooft doesn't yet fully believe in superdeterminism.

There are lots of holes, he admits — although in Vienna he did tell me that it's the only explanation that he trusts. 'I can't help being disgusted by the Many Worlds interpretation,' he said. 'I want to know what is really going on …'

I'd tell you to make up your own mind. But, if 't Hooft's suspicions are correct, you really can't.

ψ

'I like this idea,' Jerome says.

Of course he does. If it's right, nothing — his incarceration, the fate of his sons and his wife — is his fault. Everything is outside of his control. I'm tempted to point out that it also means his successes are nothing to glory in. His return from England, for example, gave him enormous satisfaction. He travelled through Europe a celebrated man. He effectively went on tour, invited to visit and be entertained by the most famous doctors, publishers, bishops, and philosophers of the continent in Bruges, Ghent, Brussels, Louvain, Michelin, and Antwerp. The procession continued to Aix-La-Chapelle, Cologne, Mainz, Worms, Speyer, Strasbourg, and Basel. From there he went on to Berne and Besançon, Zurich, and — finally — Milan.

Crowds had lined the streets of Milan in welcome, in stark contrast to Jerome's long-forgotten return from Gallarate as a penniless pauper with a wife, a young baby, and a trolley of books. 'Now they put up bowers for me to pass under and my treasure-boxes are great enough to be drawn on a cart flanked by guardsmen,' he wrote of the occasion. 'Where I was despised

I am now the chief physician, where I was unknown my skills in medicine, numbers, the stars, wisdom and machines are everywhere discussed. Honours and awards have been flung at me till I grow weary of them. Sovereigns of church, palace and battlefield have summoned me to their aid.'

Such celebrated success must have been particularly galling for Nicolo Tartaglia, a man whose reputation had, thanks to Jerome, been in tatters for some years.

Chapter 10

Let's skip back a little, to 10 August 1548. For the three years that have passed since the publication of *The Great Art*, an ever-more incensed Tartaglia has been trying to goad Jerome into a public debate. Jerome has refused every request, reasoning that he has nothing to gain. Jerome's student, the hot-headed Lodovico Ferrari could not say the same, though, and so has repeatedly challenged Tartaglia himself. Tartaglia has steadfastly refused the young upstart; what could he gain from defeating such a nobody?

This impasse continues until Tartaglia comes across the opportunity of a lifetime: a lectureship in his hometown of Brescia. He applies for the post and is told it is his — on one condition. The faculty are aware of his feud with Jerome and Ferrari — everyone is — and have decided to have some sport. All Tartaglia has to do to get the job, they say, is to beat Lodovico Ferrari in a mathematical contest.

Everything in the Garden of the Frati Zoccolanti in Milan is ready for the duel. The arrangements are spectacular. Don Ferrante di Gonzaga, the governor of Milan, takes centre stage on the dais, presiding over the day's festivities. He is flanked by everyone who is anyone in Milan — scholars, city officers, and the noblemen of the district, along with their families. Mathematics has never been so glamorous.

The church is looking beautiful and so is the garden. The carpenters have been hard at work. At the front of the stage are two desks, set facing each other. Behind each desk there is tiered seating for each man's supporters. Between them, an arrangement of flowers is spread over the grass, a symbol of the sweet atmosphere that this scholarly tussle is meant to maintain.

Each competitor will answer sixty-two problems, scientific and mathematical, posed by his opponent. Ferrari, for instance, has asked:

Divide eight into two parts such that their product multiplied by their difference comes to as much as possible, proving everything.

and:

There is a cube such that its sides and its surfaces added together are equal to the proportional quantity between the said cube and one of its faces. What is the size of the cube?

The methods for solution of the problems could be derived from the algebra laid out in *The Great Art.* But that required some deep thinking, not all of which had been completed. Take this one, for instance:

There is a right-angled triangle, such that when the perpendicular is drawn, one of the sides with the opposite part of the base makes 30, and the other side with the other part makes 28. What is the length of one of the sides?

The Great Art contained an example of such a problem, but not a general rule for solving them. This latter problem even defeated Ferrari, as it turned out. According to the rules, that means Ferrari shouldn't have set it. As Tartaglia points out. 'It is a very shameful thing,' he says, 'to put forward such a question in public, and not to know how to solve it by a general rule. I have the same opinion of your Problems 26 and 19, but I reserved to myself my reply to you on them, in front of the referees ...'

In front of the referees, it turns out, Tartaglia is himself somewhat shameful. He is irritable and poorly behaved. As the competition degenerates into a rowdy war of words, Tartaglia is repeatedly urged to 'resume the dignified mantle of the scholar' by the officials. By the end of the first scheduled day of the contest, Tartaglia has had enough. It is clear to him that Ferrari has worked harder at algebra than he, and has a better command of it. Some of Ferrari's problems, he knows, are beyond his abilities. That night, Tartaglia leaves Milan. In the morning, Ferrari is announced as the victor.

The result is a tonic to the careers of both Jerome and Ferrari. The latter is offered a plum position as tutor to the son of Holy Roman Emperor Charles V. It is a testament to Ferrari's hot-headedness that he turns the job down, protesting that it is too poorly paid. Clearly impressed by the young man's chutzpah, the

Emperor improves his offer. Ferrari can, if he wishes, become tax assessor for Milan, one of the most lucrative positions in the Emperor's gift. Decades later, Ferrari retires a wealthy man — though it is an enforced retirement. The position had involved a great deal of horse riding, which eventually gave Ferrari a career-ending fistula in one buttock.

There is a further sting in the tale of Lodovico Ferrari. Though he had risen from servant boy to Milan's chief tax collector, he had not forgotten his family, and had brought his beloved sister, Maddalena, to live with him. However, she betrayed his generosity — and was herself betrayed.

In 1560, Maddalena poisoned Ferrari with white arsenic, inherited his riches, and refused to grieve — or so Jerome records it. Fifteen days after her brother's death, she married a man who took possession of everything she owned. Within a short time he had abandoned her. Maddalena ended her life in abject poverty in a rural hovel. Jerome, in a break from his typical character, showed her not the tiniest indication of pity.

'You cannot betray your family and expect to be pitied.' Jerome's expression is steel.

'Even with a family like yours?'

Jerome stares at me, but he does not respond.

There was a time, as a young man newly qualified in medicine, when Jerome thought he would never father children. He made his doubts public — he didn't balk at discussing sexual matters, especially his impotence. In *On Subtlety*, he notes that 'it is a help to associate a great deal with pretty girls, and to read an erotic tale,' but admits that he has on occasion spent three whole nights with a girl, making 'energetic, if ineffectual, efforts to relieve his condition,' as one biographer tactfully put it.

Eating scrapings of bull penis on an empty stomach didn't help, as he had been assured it would. 'I have not yet tested whether an oven-dried wolf penis, if chewed, can instantly kindle sexual desire and the power of having sexual intercourse,' he writes. Clearly, he talks about the issue with others. He knows of a man who 'would not erect unless he was being beaten, and many who would not unless they were giving a beating'. Others need to suck on a nipple to achieve orgasm. He tells of a man who used a herb imported from the Indus: 'when he had chewed it, he could complete sexual intercourse seven times in a day'. But Jerome didn't have this herb and didn't know what might cure his impotence.

He hadn't always had a problem. Some of his writings intimate that he'd had sex successfully with a girl when he was eighteen and had liaisons involving cunnilingus and fellatio — with males and females — before that. He writes about enjoying erotica and frequent masturbation. But, by the time he complains about his impotence in a letter to a friend, it has clearly been going on for some time. 'I maintain that this misfortune is to me the worst of evils,' he says. 'I bitterly weep this misery, that I must needs be a laughing-stock, that marriage must be denied me, and that I must ever live in solitude.' He is unable to explain the cause, but laments that it means he eventually shunned the society of women, bringing on himself suspicions of 'still more nefarious practices'.

Nonetheless, the advent of the angelic Lucia cured him and he did have children: first Giovanni, then his daughter Chiara, and finally Aldo. All three brought him a share of sorrow.

Giovanni was born to a bad portent, with his curved spine, his webbed toes, and the wasp at his baptism. No wonder that, when Chiara is born, Jerome remarks with relief that the horrors of Giovanni's early years were not repeated: 'She was in no way disfigured as was our firstborn, nor was her baptism marked by any untoward incident.' Her only bad luck was to be unable to conceive children. This was, in time, a severe blow to Jerome, who craved grandchildren. Nor did Giovanni or Aldo have children — but that was the very least of the problems they brought him.

For starters, Giovanni is quite simply a dullard. Somehow, with heroic effort and no small amount of string pulling, Jerome manages to push him through medical school. He then uses his influence to have Giovanni admitted to the Milanese College of Physicians. His son will make a hopeless doctor, Jerome knows, but even he doesn't see just how hopeless. Soon after his practice begins, Giovanni inadvertently administers a fatal dose of poison to a local official. 'That was, indeed, the beginning of all ills,' is Jerome's later description.

There is no proof that Giovanni meant to kill the clerk, but that is hardly the point. What matters is that details of the deed have fallen into the hands of the Seroni family.

The Seronis were once wealthy. However, the father, Evangelista, has spent every crown. Destitute and desperate, they have turned to crime. They have never been known to miss an opportunity for blackmail — it is something of a speciality — and Giovanni is an opportunity on a silver plate. The silverware in this instance belongs, as do many other desirable things, to Jerome. A proposition comes to Giovanni: you can have your crime exposed to the College of Physicians, or you can marry our Brandonia. Giovanni must have had to think hard. Brandonia Seroni is fat, plain, bad tempered, and widely known to be promiscuous. She is, as Jerome put it, 'destitute of all good qualities': marriage material only for someone who had no option. Giovanni accepts the proposal.

Jerome only learns of the marriage when it is a *fait accompli*. A breathless servant who had been looking for his master pants out the news: Giovanni has come home with Brandonia as his wife and is looking to set up a household under Jerome's roof. That household, Jerome quickly learned, would include most of the Seroni clan. He can see that the Seronis' plan is to bleed the wealthy doctor of all his money and so Jerome refuses to let Brandonia in the house. Giovanni is furious — and no doubt terrified at what might happen to him if the Seroni plot is foiled. His father refuses to back down, but writes him two books' worth of advice — he calls the volumes *Consolation* and *Adversity* — and offers some financial support. His letters chastise Giovanni for bringing all this trouble upon himself: 'for all the ills that now hang over you, your poverty, your wife, your ill repute, your absence from your father's house, all these I say you have prepared for yourself willingly and knowingly'.

Perhaps that judgement is what drives Giovanni to kill his wife.

ψ

Jerome is in his university lodgings at Pavia when the letter from the rector of the Milanese College of Physicians arrives. His spidery scrawl states simply that Giovanni, his younger brother Aldo, and a servant have all been arrested in Milan for the murder of Brandonia Seroni. Jerome is to return to Milan immediately, the rector says. The good name of the College must be defended.

'He said nothing about defending my sons' names — nor my own,' Jerome says.

'Did you ever doubt they were innocent?'

Jerome gives me a long stare. 'I never doubted Giovanni,' he says. 'My first thought was that Aldo must be to blame.'

Despite what I have read, I am still shocked at Jerome's matter-of-fact tone. 'You never thought Aldo might also be innocent?'

Jerome shrugs. 'You don't know my younger son, do you?'

Chapter 11

Aldo was born on 25 May 1543. There were no evil omens. 'A fine child,' Jerome said, 'with no disfigurement or fault attending him, or any trouble to my wife.' Perhaps that should have taught Jerome the fallibility of his 'signs and portents' approach to predicting the future. Just over three years later, Lucia, Jerome's wife, was dead and the boy was being raised by a nursemaid.

In 1546, Lucia had slipped into sickness. Jerome mentions a 'decline', a 'paleness' and 'lassitude', but little else. Towards the end of that year, she died, aged thirty-three. Lucia's passing seemed to happen in the background. Jerome writes more about the syphilis-ridden genitals of his friend Ottaviano Scoto than he did about the illness that robbed him of the love of his life.

Perhaps some things are too painful to dwell upon. 'She was brave, indomitable in spirit, gentle, affectionate, fine to look upon,' he wrote. And that was it — as if he could bear to write no more. She had captured his affection for sixteen years, suffered poverty and loss, seen her jewels gambled away, enjoyed a few years of prosperity, and then, aged thirty-three, she was gone. Jerome was left alone with his children.

Aldo grew up in the care of a servant and he did not grow up well. He was brutal and cruel to the household animals, of whom there were plenty — Jerome is fond of creatures. In his early teens, Aldo developed a habit with the dice, but had none of his father's skill with numbers. When the addiction raged, it turned him towards the life of a criminal. First he stole from his father's household and then, when possessions began to be more securely locked away at home, he robbed others. His father despaired at the money he spent bribing the authorities to let Aldo out of jail: 'thousands of crowns to release him from just punishment'.

Yet if Jerome found things missing from the house, it can't have given him any less grief to find the things Aldo occasionally left strewn about the place. They were receipts for work carried out, but not the kind of work to warm a father's heart. 'Messr Aldo Cardano, public executioner, for torturing by rack and vice, Valentino Zuccaro, 3 scudi. And to the same for having in due course burnt Zuccaro and thrown the cinders of his flesh into the river, a further 7 scudi.'

Aldo is now a freelance torturer, working — when the opportunity arose — for the Inquisition.

$$\psi$$

Jerome is to hurry back to Milan, the rector writes. Jerome must put every effort into supporting his sons' exoneration, 'that their innocence may be defended with that of the name of this learned College, lest it fall into disrepute by association'. The letter is blatantly self-serving and the irony is not lost on Jerome. He had barely been elected to the College because of the shame of his bastardy. Suddenly, when the growth of his fame has brought the College a reflected glory, he is the one called to defend its reputation; 'Now, bastard or no, I was commanded to still the fluttering in the dovecotes of respectability,' he later recalled. There is no wry smile here, for Jerome is utterly wretched at the woeful predicament of his children. 'I cared no more for my credit in Milan or the world,' he wrote later. 'I cared only to throw all my influence at the feet of my sons.' And so he sets out, determined to defend them. 'A man sixty years old, greyheaded and bent, is no less able to plead with judges,' he says.

The crime, as laid out by the prosecution, is the poisoning of Brandonia Seroni by arsenic that was baked into a cake. But no fewer than five doctors from the self-serving Milanese College of Physicians are on hand to testify that Brandonia had not died by arsenic poisoning; the appearance of the corpse is all wrong, if these fine upstanding members of society are to be believed. Arsenic poisoning blackens the tongue, they say; it lifts the fingernails and corrodes the internal organs. None of this is true of Brandonia's body, they report. In their learned conclusion, Brandonia had been suffering from lipyria, a wasting disease. This, they insist, is the true cause of death.

The Seroni family are having none of it. With a flourish, their lawyers bring out the servant who baked the cake. He overheard Aldo and Giovanni plotting, he says. He was handed a vial of liquid to put in the cake mixture, he had baked the cake and served it to his mistress; she vomited immediately.

The doctors counter that Giovanni, being a qualified doctor, must have intended the mixture as a treatment for some malady his wife was suffering and that such treatments do occasionally cause accidental harm. White arsenic, they point out, is a treatment for lipyria. Perhaps Giovanni had simply erred in the dosage?

A stalemate ensues. Then, several days into the trial — out of the blue — Giovanni confesses to premeditated murder. This, he announces, was his third attempt to kill his wife; the first two times he tried, the poison failed to do its job.

Until this point, Jerome has not been allowed to speak in his son's defence because the law does not permit relatives to offer evidence. Now, with the court debating only motive, Jerome is permitted to address the assembly. What a shame, then, that oratory was not his gift and that, aged twelve, he had abandoned the study of persuasion.

As a young boy, Jerome had decided to follow a path of medicine and mathematics, rather than the law. 'I had had enough and enough of law; its books, by my experience, were heavy and its conclusions unsatisfactory ... numbers I thought greatly of for what they could prove.' At this juncture in the trial, the old man, now in his fifty-ninth year, must have suffered a twinge of regret about that decision because he turned out to be a poor advocate for his son.

Even by the standards of the day, Jerome's two hour speech is odd. He argues that poisoning is barely murder — at least the boy hadn't *stabbed* his wife, which would have been far worse. Then he addresses the character of the victim. Brandonia was no saint, as everybody knew. Jerome rakes through the dead woman's list of crimes without mercy or respect for the deceased. She was an unfit mother who had let her first child die through neglect. She had given birth to two more children during the marriage whom Jerome, worried for their safety, was now raising. What's more, she had openly told Giovanni during the course of a heated row that neither child was his. Her mother, who was in the room at the time, backed up Brandonia and went so far as to name the fathers who had cuckolded Jerome's son. Such provocation, he argues, almost justifies his son's crime. His speech was a character assassination that must have stirred venom in the hearts of her family.

And then comes the condemnation of Aldo. Giovanni, Jerome tells the court, is just too simple-minded to have hatched such a plot. He must have been under the influence of his wicked younger brother, who probably persuaded him to confess, too. 'Surely my son is worthy of excuse and pardon,' Jerome cries to the court in his impassioned summation. 'A youth as simple of wit as any in the state ... He is so simple that I take no more thought in the buying of my shoes than he took in the marrying of his wife ... Was he not foolish, if he meant murder, in choosing as his confidants a mischievous brother and a servant boy who would break any silence for reward ... Would you sacred senators inflict death on a lunatic who in a lucid interval killed a man?'

To be fair to Jerome, his sons had, up to that point, served as each other's accomplices. They had worked together to strip their father's house of saleable assets and no doubt had split the proceeds. Aldo had to pay off his gambling debts to avoid being murdered by his benefactors. Giovanni, if he could not take his in-laws into the house, would take the house — or at least its contents — to his in-laws. Brandonia had even handed her father her wedding ring, a gift that Jerome had made to his son, so that her father could sell it.

It is worth noting that, by the time Jerome accuses Aldo, the old man is altogether wrung out. He has spent all his money on Giovanni's defence. He has argued for two hours in his elder son's favour. He is tired and emotional.

Somewhat oddly — and possibly self-defeatingly — Jerome then switches to lauding Giovanni's academic achievements: 'Is he not a baccalaureat, a man honoured by Academy and College … He is a learned man in his profession. Is the head matured and educated by so many nights of toil to be cut off like the head of a man as ignorant of yesterday as of tomorrow?'

His closing words, tears running down his face, are a straightforward plea for his son's life. 'O Sacred senators, you cannot condemn a son to the galleys without condemning to a worse fate the father, who is innocent; and to kill him would be a fate worse than death to me. I beseech you therefore, that if you prove him guilty you sentence him to perpetual exile, and spare him his life and dignity. For in that way you will also spare mine.'

The pleas fall on deaf ears. The Senate sentences Giovanni to death. The only way out, the court says, is if the Seronis could be satisfied with financial compensation — and they are to name their price. 'They demanded,' Jerome recalls, 'more gold and treasure than could be found in the coffers of a king.' They know he can't meet their demands — for once, money came second. They are far more interested in humiliating Jerome and witnessing the execution of his worthless son.

As it turned out, there were no public witnesses. Jerome's twenty-six-year-old son was executed within the prison walls that same night. On 8 April 1560, Jerome takes delivery of his son's decapitated body. Brandonia's daughter dies in the same week, quickly followed by the nurse whom Jerome had hired to look after his daughter-in-law's children.

In the space of seven days, the broken old man has arranged and funded three funerals.

'It must have been appalling. I can't imagine how that feels.'

I am sat on one end of Jerome's straw mattress; he is lying, curled up in a ball, at the other. He doesn't respond straight away; he has told me the story of his son's demise. Even though those events unfolded a full decade ago, the revisiting of it has exhausted him. I sit in silence. After ten minutes, he pulls himself upright and, hands holding the green jewel to his lips, turns his head towards me.

'Do you have children?' he says.

'Two,' I reply. 'A girl and a boy.'

'How would you feel to lose one?'

'I think it would crush me,' I say. 'I don't think I would ever recover.'

'I am certain I will not,' he says.

ψ

Certainty is always misguided. It is something we use to comfort ourselves, a delusion we often indulge in the wake of disaster. We become certain 'some good will come of this', or — as in Jerome's case — that recovery is impossible. Perhaps it says something about humans that we never feel certain about anything when life is going well. A state of happiness is rarely taken for granted. That, perhaps, is when we are at our most perceptive because the appalling truth is that uncertainty is inherent to the cosmos.

There is a chapter in Jerome's autobiography called 'The Disasters of My Sons'. Within its few short pages, Jerome describes Giovanni's calamitous marriage and eventual execution, and 'the folly, the ignominious conduct, and violent actions' of his younger son, Aldo. He is not seeking pity — Jerome writes that he is 'by no means unaware that these afflictions may seem meaningless to future generations, and more especially to strangers' — but wants to make a point. There is, he says, nothing in this mortal life 'except inanity, emptiness, and dream-shadows'. The only basis on which mortals can find a firm foundation for their lives is to extract wisdom from significant events. Within the great adversities of life, Jerome argues, 'mortal things may find, now here, now there, new meaning and testify that they are destined for a purpose and a use not to be despised'.

It seems like a reasonable attitude. But quantum theory tells us it is utterly mistaken. So far, we have stated that God does play dice, that there is no pattern or purpose behind the cosmic drama of existence. Now we are about to delve deeper into the mathematics behind the Schrödinger equation and discover why. The fundamental rules of the universe — rules that Jerome helped to formulate, let's remember — tell us that the universe only exists because of a random event. Even the formation of galaxies, stars, planets, and people is dependent on random chance. If quantum theory is to be believed, there is no purpose except that which we deluded beings construct for ourselves.

I am talking about quantum theory's 'uncertainty principle'. We skirted this earlier, when we talked about entanglement, but this seems a good moment to face it head on.

The uncertainty principle may be the least understood concept in physics, which is somewhat ironic. It is nothing to do with practical problems, such as error-prone measurements. It starts from the mathematics behind the Schrödinger equation, which says that multiplying a by b is not the same as multiplying b by a. That sounds ridiculous when we are used to a world in which three times five gives the same answer as five times three. But this quantum world, as we are discovering, is very different from our own.

In the Schrödinger setup, the things we seek to multiply together are not straightforward numbers, but pairs of quantities such as a particle's position (we'll call it p) and its momentum (which we'll call q, so as not to confuse it with the shorthand for mass). In this notation, multiplication is denoted by putting two things next to each other. So position times momentum is pq. And that is *not* the same as qp. Why? Because the mathematical rules that govern operations with the Schrödinger equation are not the same as standard multiplication.

The difference between pq and qp is given by a simple quantity. It involves Planck's constant (h), Jerome's imaginary square root of -1 (i) and π. In mathematical notation, that is: $pq - qp = ih/2\pi$.

There are other forms of the Schrödinger equation where different pairings follow the same rule. Energy and time, for example, make another such couple. The upshot of all of them is that it is impossible to calculate a precise value for both parts of a pair. If I am applying the equation to an atom and I want to know its position precisely, I have to sacrifice accurate knowledge of its momentum — and vice versa. The more accurately I know one of these quantities, the less accurately I know the other.

This unavoidable gap in our knowledge is *not* a consequence of some inability to make accurate measurements. It is written into the theory. And it means that you can't ever predict the future state of a quantum system. That's because you can't plug exact values for all its properties into an equation that will let you work out how its state will evolve. There will always be some uncertainty in the sum of the starting conditions, so you will always be uncertain about its future.

That said, there is a link to the practicalities of measuring these not-exactly-wave, not-exactly-particle objects. If we want to find an object's position, for instance, we have to bounce something off it — a photon of light, say. But the very act of bouncing a photon off the object will give it a kick, altering its momentum. So we have gained information about position at the cost of obscuring information about its momentum at that moment. Similarly, finding its momentum involves measurements at two different times in two different places, which means the associated position is rather vague — by the time our momentum measurement is in, the position has changed. Another source of uncertainty.

Finally, it is worth noting that the principle seems to be linked to the phenomenon of entanglement. You can use quantum mathematics to show that two entangled objects — that is, two objects that have distributed the information about themselves between the two of them — are less subject to the uncertainty principle than two objects that have no connection. Experiments have shown that the uncertainty principle applies to the first measurement on one photon of an entangled pair. But when a subsequent measurement is done on the second photon of the pair, the information gained about the state of the first one is more precise than came in the first measurement. Repeat that process, and you can know the state of the first photon to arbitrary precision.

Given that entanglement defeats our understanding of space and time, there is little point trying to make full sense of the uncertainty principle. But it does seem to be fundamental and related to issues of the amount of information carried by quantum objects and their entanglement partners.

Uncertainty and entanglement are also somehow related to the second law of thermodynamics, which says that every process in the universe tends towards producing disorder. It seems to be a fundamental, defining principle behind the way the universe operates — classically and quantumly. Imagine the position and momentum of an electron as two intertwined streams of information, each one encoded so that the more you read of one, the less you can read of the other. Essentially, it's another formulation of the uncertainty principle. However, working from the Schrödinger equation and our knowledge of thermodynamics, researchers have shown that the energy of the electron is related to the information needed to describe it, and the information uncertainty prevents you from extracting more energy than the system contains. In other words, without the uncertainty principle, we would break the second law of thermodynamics.

Nothing has ever broken the second law. The physicist Arthur Eddington once said you should never back an idea that opposes it. 'If your theory is found to be against the second law of thermodynamics I can give you no hope,' he said, 'there is nothing for it but to collapse in deepest humiliation.' Quantum theory, though, is safe.

What is this telling us? Somehow, the Schrödinger equation and its answer to de Broglie's challenge about everything having both wave-like and particle-like properties, has tapped into something that is utterly fundamental to the universe, more fundamental even than space and time. This fundamental uncertainty is more than an abstract notion or an impediment to our experiments. Its application to the energy and time associated with quantum objects affects their very existence. It means there is no such thing as completely empty space, for instance, because that would imply that the universe had precisely zero energy and nothing can have a precise value of anything. A consequence of this is that, for a short enough time, the universe will lend energy to a particle so that it can come into being. And so the emptiness is filled with a constantly appearing and disappearing set of 'virtual' particles.

These virtual particles have real, physical effects. One is known as the Casimir effect. To see this, put a pair of metal sheets next to each other in a vacuum. They will move towards each other because the virtual particles appearing in the empty space will create tiny electric fields that interact with the electrons in the metal. The geometry of those fields differs depending on whether the particles are confined between the plates or in the empty space on either side of them. The difference between those geometries means the plates feel a force that pulls them together. And so they move. The movement was first measured in 1948: it is real and so, therefore, is the fundamental uncertainty in the universe that was revealed by the Schrödinger equation.

But this fundamental uncertainty doesn't just apply to events within our universe. It applies to the universe itself. The physicist's explanation for the cosmos's existence is that it came into being because of an uncertainty in something outside the physical space of the universe. Maybe a type of Hilbert space, or Jerome's *aevum* — we don't really know. But, as in the Casimir effect, that uncertainty led to a moment of spontaneous creation. We call it the Big Bang. When we look at the evidence for the Big Bang — a sea of primordial photons known as the cosmic microwave background radiation — we see that there are occasional random fluctuations in the photons' energy. These fluctuations were the seeds for the stars and galaxies that led, ultimately, to our existence. Life, like the cosmos it inhabits, is born of randomness, and there can be no such thing as being 'destined for a purpose'.

There is no comfort to be found, Jerome.

Chapter 12

For a decade after his son's execution on 7 April 1560, Jerome's life spirals into chaos. The first three years are perhaps the gentlest on him. At first, he feels vindicated by the strange fates that befall those who condemned Giovanni. The president of the court has to feign his own death to avoid prosecution for his wife's murder. A senator drowns after falling from a bridge. Another senator contracts phthisis, coughs up a lung, and dies. Evangelista, the Seroni family's patriarch, is jailed, stripped of his position as a debt collector, and becomes a beggar. The prison governor is sacked and is similarly forced to beg on the streets. The prosecuting counsel loses a son to smallpox and a jurisconsult who gathered evidence against the Cardanos is jailed for corruption. 'Of all those who brought accusation against my son not one escaped without some terrible calamity — being either smitten or destroyed,' Jerome recalls.

Soon after all this, perhaps because of it, Jerome is greeted everywhere with suspicion. He still holds academic positions in Milan and Pavia, but he has become someone to avoid. 'As I walked about the city men looked askance at me; and whenever I might be forced to exchange words with anyone, I felt that I was a disgraced man.' Aware of the atmosphere, he withdraws from society. 'I had no notion what I should do, or where I should go. I cannot say whether I was more wretched in myself than I was odious to my fellows.'

Jerome is soon convinced everyone in Milan is plotting to kill him. A return to Pavia does nothing to relieve his fears. He becomes a wanderer. It is a time he will choose not to look back upon when he writes his memoirs. Others make mention of seeing him in Padua, Milan, Bologna, and Pavia, but Jerome leaves no record of what he did in those cities or what impels him to move between them. He writes: terrible ravings, most of them, philosophical volumes strewn with nonsensical rants and asides. But there are gems, too, in the form of geometry, philosophical, and medical textbooks that will be the standard texts for decades to come. His readers will include Johannes Kepler, Johann Wolfgang von Goethe, and William Shakespeare. Scholars of the Bard have even identified an English translation of *Consolation*, Jerome's three volume meditation on dealing with tragedy and disappointment, as the book Hamlet carries during the soliloquy that begins 'To be or not to be ...'

POLONIUS: What do you read, my lord?
HAMLET: Words, words, words.

What words? 'In holy scripture, death is not accounted other than sleep, and to die is said to sleep,' says Jerome in *Consolation*.

'Seeing, therefore, with such ease men die, what should we account of death to be resembled to anything better than sleep...' Hamlet, for his part, says, 'To die, to sleep / — no more — and by a sleep to say we end / The heartache and the thousand natural shocks / That flesh is heir to ...'

I could cite more examples, none entirely conclusive. But in 1807, Francis Douce, Keeper of Manuscripts in the British Museum, said, 'Whoever will take the trouble of reading the whole of Cardanus as translated by Bedingfield will soon be convinced that it had been perused by Shakspeare.' Six years later, Joseph Hunter, author of *New Illustrations of the Life, Studies, and Writings of Shakespeare*, cited passages of Jerome's *Consolation* that 'seem to approach so near to the thoughts of Hamlet that we can hardly doubt that they were in the Poet's mind when he put [certain speeches] into the mouth of his hero'.

Anyway, all that is in the future. In the present, Jerome must live with a debilitating paranoia. Wooden beams and lead weights, he believes, have been balanced above doors so as to fall on his head. At mass, he thinks that the choirboys are whispering plans to poison the wicked old doctor. This, he remembers, for he writes it in a chapter of his autobiography entitled 'Perils, Accidents, and Manifold, Diverse, and Persistent Treacheries.' In the end he outlives all his would-be assailants: 'all who sought my life perished,' he says. He cannot avoid calamity forever, though.

As 1563 draws to a close, the Milanese College of Physicians withdraws his right to teach. Jerome is accused, by numerous witnesses and in sworn statements, of sodomy and incest. There is no trial, just a sentence. Now, aged sixty-three, he is an exile from Milan. 'Reduced once more to rags, my fortune gone, my income ceased, my rents withheld, my books impounded, my only companions prejudice and calumny,' he writes. He resides in a Padua workhouse for a time. He is permitted to treat plague victims at a monastery in nearby Gallarate, but no other work is forthcoming — and the shadow of the Inquisition is looming larger. Across Europe, thinkers and writers are being arrested, questioned, and sometimes tortured and executed in an attempt to quell the growing tide of dissent against the Church. These days, if you have published novel, provocative ideas, you can be sure they will have been read by the Inquisitors. And if you are without friends or money, you are unlikely to survive their attention unscathed.

ψ

The spirit of the Inquisition has never been fully extinguished, as David Bohm would testify if he were still among us. During the latter part of World War II, when he was a graduate student at the University of California, Berkeley, Bohm's PhD supervisor, J Robert Oppenheimer, recruited him into the newly formed effort to build an atomic bomb. Bohm's contributions to the Manhattan Project were so valuable that they were immediately classified and Bohm was shut out, not even being allowed to write his own PhD thesis. He did get his PhD, though, after insisting that Oppenheimer vouch for the quality of his work.

By 1950, Bohm was working with Einstein at Princeton, where his past came back to haunt him. Early in his PhD studies he had joined a trade union and, briefly, a couple of communist groups.

Those communist associations, coupled with the national security implications of his PhD work, made him a target for Senator Joe McCarthy's crusade against un-American activities.

Bohm refused to answer questions, and refused to name anyone that the McCarthyists should investigate. He was arrested. By the time he was acquitted, he had been suspended from Princeton. In 1951, unemployable in the United States, Bohm took a job in Brazil. The US authorities then confiscated his passport and he was forced to apply for Brazilian citizenship. It was as a Brazilian that he travelled to England and began a long career as a professor of theoretical physics at Birkbeck College in London. There, he successfully applied for a British passport. Then, in 1986, he won back his American citizenship in a legal battle with the US government.

Nothing in that long and painful saga distracted David Bohm from physics. He made significant contributions in a variety of areas, but it is for his interpretation of quantum physics that he is best known. In 1952, Bohm published a seminal paper that is now seen as a complementary, but independently derived, version of work begun decades before — and then abandoned — by Louis de Broglie.

Let's go back to the experiment where we fire those quantum arrows through those slits and get a weird pattern. While the Copenhagenists would say the arrows have no definite position or momentum until they hit the target, de Broglie formulated another idea, written up in his 1924 dissertation. He brought it up again when he gave a talk in October 1927, at the same meeting where Einstein and Bohr had their famous debates over quantum theory. In his talk, he spoke about the '*théorie de l'onde pilote*' — pilot wave theory.

According to de Broglie, each photon fired at the double slit exists as a real object. He suggested it has a definite position and momentum at all times. What you can't know is the initial position. And since the initial position would be what you combine with the momentum to give you the final position, you can't know the final position in advance, explaining the apparently random outcomes of each measurement.

Because it is a real object, with a well-defined position, the photon can pass through only one of the slits. However, its trajectory is guided by a 'pilot wave', in much the same way that a ferry entering a treacherous harbour is guided by a pilot boat. This pilot wave is also real and has properties that are a reflection of the wave function in the Schrödinger equation.

Because of this link to the Schrödinger equation's wave function, although the particle will only pass through one of the slits, there is still a final distribution of particles determined by an interfering wave. That means the major consequence of interference — the strange clumping at certain points on the target and absence at others — will occur.

Eventually, de Broglie abandoned his idea and became a Copenhagenist. It wasn't that the pilot wave theory was particularly flawed; it was just that Bohr was probably too powerful and charismatic a figure to resist. So the pilot wave theory sank.

In 1952, however, it resurfaced in the hands of David Bohm. Bohm's idea of an invisible, undetectable pilot wave was roundly criticised, but a man who had survived the McCarthy witch hunts was not easily put off. Having overcome the most heinous character assassination of the era, he could take a little heat. And so he stuck to his guns, suggesting we needed to look at quantum experiments in a different way. In a 1952 paper, published in *Physical Review*, he said, 'the history of scientific research is full of examples in which it was very fruitful indeed to assume that certain objects or elements might be real, long before any procedures were known which would permit them to be observed directly.' In other words, why shouldn't there be an as-yet-undiscovered pilot wave?

> *Of course, we must avoid postulating a new element for each new phenomenon. But an equally serious mistake is to admit into the theory only those elements which can now be observed ... In fact, the better a theory is able to suggest the need for new kinds of observations and to predict their results correctly, the more confidence we have that this theory is likely to be good representation of the actual properties of matter and not simply an empirical system especially chosen in such a way as to correlate a group of already known facts.*

So far, so good, perhaps. But there are two problems. The first is that, in order to get the predictions right about the interference effect and the ultimate distribution of the photons at the detector, you have to work backwards from the final result.

The second problem is that Bohm's pilot wave is odd — in a way that physicists call 'nonlocal'. This means that the properties and future state of our photon are not determined solely by the conditions and actions in its immediate vicinity. The photon's pilot wave and the photon's wave function are linked to the wave function of the much, much larger system in which they sit — the wave function of the whole universe, effectively. So our photon can be instantaneously affected by something that happens half a universe away.

Many physicists — most physicists — are not happy about allowing this nonlocal action. After all, such action is prohibited by Einstein's special theory of relativity, which says an influence can't travel faster than the speed of light.

On the plus side, it does give us an explanation for entanglement-based phenomena. And it's not clear that accepting Bohmian mechanics is any worse than shoehorning entanglement into a relativity-friendly physics. Many fine physicists are certainly happy to talk in terms of Bohmian mechanics. In Vienna, for instance, an experimenter called Aephraim Steinberg explained his experimental results from a Bohm-eyed view; this, he says, is the easiest way to think about it. What Steinberg presented was a picture showing the trajectories of photons as they pass through the double slit apparatus. In the Copenhagen interpretation, remember, this is impossible because the photons have no meaningful existence before they are detected. Without an existence, they can't logically have a trajectory.

So how did Steinberg come up with the photons' trajectories? The answer is, by using Yakir Aharonov's weak measurement. All things are indeed connected.

The de Broglie-Bohm interpretation of quantum physics, as it is now known, is not popular. Only one venerated physicist has

ever really championed it: John Bell, the Irishman who came up
with the test for the existence of entanglement. Here's what Bell
had to say:

> *While the founding fathers agonized over the question*
> *'particle' or 'wave', de Broglie in 1925 proposed the obvious*
> *answer 'particle' and 'wave'. Is it not clear from the*
> *smallness of the scintillation on the screen that we have to*
> *do with a particle? And is it not clear, from the diffraction*
> *and interference patterns, that the motion of the particle is*
> *directed by a wave? De Broglie showed in detail how the*
> *motion of a particle, passing through just one of two holes*
> *in screen, could be influenced by waves propagating through*
> *both holes. And so influenced that the particle does not go*
> *where the waves cancel out, but is attracted to where they*
> *cooperate. This idea seems to me so natural and simple, to*
> *resolve the wave-particle dilemma in such a clear and*
> *ordinary way, that it is a great mystery to me that it was*
> *so generally ignored.*

Bell felt de Broglie-Bohm was a better bet than anything
the Copenhagenists had to offer. They had elevated the issue
of measurement to the status where it was fundamental to the
subject without ever making clear what it actually entailed. 'The
concept of "measurement" becomes so fuzzy on reflection,' Bell
said, 'that it is quite surprising to have it appearing in physical
theory at the most fundamental level ... does not any analysis
of measurement require concepts more fundamental than

measurement? And should not the fundamental theory be about these more fundamental concepts?'

Bell is widely venerated. Go to quantum physics conferences and his name comes up again and again, with some people quoting from his writings as if from scripture. He has the advantage, from the fame perspective, of having died suddenly and relatively young. A cerebral haemorrhage took him out of the blue in 1990, aged just sixty-two. But even his influence is not enough. When it comes to quantum interpretations, the Copenhagenists appear to have won the day. How? By sheer weight of personality.

Niels Bohr, in particular, was so influential that he controlled much of the funding available for quantum research. He was also a likeable character: people enjoyed his company, craved his approval, and tended to bow to his way of thinking. Too little is made of the importance of personality in science. Some have called Bohr a bully, but unfairly, I think: he was simply persistent in arguments and reluctant to change his mind. In one lengthy discussion he reduced Werner Heisenberg to tears. In another, Schrödinger fell ill while staying at Bohr's house, took to his bed, but was still harangued by his host, who sat on the edge of the bed and continued their argument.

ψ

'One man can change everything,' Jerome says, looking around his cell.

I assume he is thinking of Nicolo Tartaglia. 'Yes,' I say. 'It is astonishing what The Stammerer has done to you.' My eyes burn through the gloom of the cell, aching with pity for poor Jerome. Tartaglia, his every action toxic with malice, has arranged for the full weight of the Holy Church to fall on Jerome. 'He has finally wreaked his revenge.'

Jerome looks at me, puzzled. 'I don't think so,' he says.

My eyes narrow in protest. 'Yes. It was Tartaglia who alerted the authorities, who oversaw your arrest. With your son. Aldo traded your freedom for a post with the Inquisition.'

'What makes you think that?'

'I read it. In a book by a journalist called Alan Wykes.'

Jerome shakes his head. 'I can believe it of Aldo,' he says. 'Aldo would do all you suggest, I agree.' Jerome's eyes are narrow, staring intently toward me as if attempting to diagnose my mental state. I can feel that something is wrong.

'But?'

'But Tartaglia has been dead for more than a decade.'

I don't know what to say.

Chapter 13

I'm so embarrassed. How could I have fallen for Wykes's fabrication? For all my supposed knowledge of the universe, Jerome has exposed my utter foolishness. Surely Wykes hasn't just shifted the date of Tartaglia's death to fit with his own desire for a satisfying narrative? I have no choice but to dig for answers. Alan Wykes wrote a slew of books. Would anyone have kept his notes? I can't help thinking of my own family; it seems a long shot that they would keep mine after I'm gone.

An internet search tells me many things. Wykes died in 1993 after a long career as an 'author, journalist, raconteur and professional clubman'. He was 'a prolific storyteller with a prodigious memory for historical detail'. But people did question his reliability. A book on the composer Lord Berners had people grasping for Wykes's sources for certain anecdotes: it turns out he learned the stories at second or third hand and they remained uncorroborated. Similarly, his volume on Adolf Hitler contained the contention that the Führer's anti-Semitism arose when he contracted syphilis from a Jewish prostitute — an idea roundly demolished in previous studies. 'Wykes's witnesses were either mistaken or dishonest,' one reviewer says. My suspicion is that the witnesses may not have existed.

I can find no mention of a surviving spouse or children. But there is the 'clubman' angle. For more than a quarter of a century, Wykes had been honorary secretary and chairman of the benevolent fund of the Savage Club, a London gentleman's club whose former members included Sir Edward Elgar, Dylan Thomas, and JM Barrie. The internet is a goldmine here too: Mark Twain had once been a guest at the club; Sir Arthur Conan Doyle mentions the Savage Club in *The Lost World*; and the fascist Oswald Mosley turned up one evening (as a guest of Henry Williamson, author of *Tarka the Otter*), but was immediately asked to leave. I email the club. And then, a couple of weeks later, my phone rings and I find myself speaking to Philip Voke, the man who inherited all of Wykes's notes and manuscripts.

The two men had met in the 1970s, when Voke moved into the house next door to Wykes. The boxes of materials left to Voke after Wykes's death are now in the possession of Reading Library. So, within a couple of weeks, I have lunched with Voke at the Savage Club and taken a train to Reading.

It is a short walk from Reading station to the Library's reading room. Adrenaline is pumping through my body as I collect the box I have requested. I lay out my notebook and pencils — no pens are allowed in the reading room — and open the box.

In the movie of this book, there should be a quiet, apprehensive moment of tight focus on the box, with perhaps a single instrument sounding out the anticipation in a low tone. But there is no

Eureka! moment to merit such drama. I do not unearth a pile of missing notes and sources, and find everything suddenly making sense. The problem is that the footnotes and references in Wykes's handwritten version of his Doctor Cardano book are the same as in the typescript, which are the same as in the published version — with just one exception.

The note '2' on page 174 of the published hardback is not in the manuscript or the typescript. This is the crucial fact. The published book, where the note does appear, has no corresponding reference. Had it been there, it would have referred to the passage where Wykes details Aldo and Tartaglia's direct involvement in Jerome's arrest:

> *On 13th October 1570, at Tartaglia's instance, Cardano was arrested while in Bologna. 'The boy Aldo, to whom I had promised the reward of the appointment of public torturer and executioner in that city, came to me in Rome with the intelligence that his father was in Bologna awaiting an interview with the syndics. I thought to myself, "Ah! This will be pleasant, to raise his hopes that at last the restrictions are to be lifted from him and then, an instant before the realisation of those hopes, to cast him into prison." And so it was. I hastened to Bologna, and there he still sheltered, in the ruins of a hovel, awaiting an ascent to his former status. I instructed the guards to arrest him as he set out for his appointment.' [2]*

I can come to only one reluctant conclusion. It is all made up. At the last minute, perhaps, someone at the publishing house called Wykes and asked that he declare his source. They inserted the '[2]' in anticipation of his supplying some reference. He never did. How could he? At this point, Tartaglia is long dead, his body lying in the crypt of Venice's San Silvestro church.

Within the box are some pages of handwritten notes that look like an author's snagging list. Some are struck through in red ink. On one page I read 'Aldo — what happened to? Chiara ditto' and 'Aldo part of plot with T?'.

In the synopsis of chapters, written before embarking on the manuscript, the facts are laid out unembellished. Wykes has written to his publisher, Frederick Muller Ltd, that he aims to 'produce a fully documented piece of research that would, I hope, be accepted as a serious contribution to historical literature'. There is nothing here about Jerome's daughter Chiara, whom Wykes's published book painted as a brazen nymphomaniac. In his narrative, she seduced her older brother and rendered herself infertile when she aborted Giovanni's child. Wykes has Chiara's husband complain to Jerome that she has 'insatiable lusts' and carries incurable syphilis that has withered away her charms: 'she is like a dead twig and bears nothing but her pudendum'. Neither is there any mention of Aldo. Even Tartaglia doesn't merit a mention in the synopsis. Somewhere along the path to publication, Alan Wykes succumbed to an ever-present temptation for writers. He crafted a story that was too good to be true.

'He wasn't the first.'

Jerome seems almost pleased. I am crestfallen to be found

wanting in my research, and he is trying not to smile. Perhaps it fuels that insatiable desire for eternal fame: he clearly enjoys being the focus of books written hundreds of years after his death. And Jerome is right. The fabrications began even before his arrest.

In 1557, the year in which Nicolo Tartaglia died, an aspiring intellectual named Julius Caesar Scaliger published what has been called 'the most savage book review in the bitter annals of literary invective'.

Exercitationes comprised a nine hundred page refutation of Jerome's book *On Subtlety*. To be fair, Jerome's work is a ridiculously wide-ranging book. Within its thousand pages he discusses the structure of the universe; lactating males (Jerome describes one Antonius Benzes, whom he met in Genoa: 'thirty-four years old, pale, with a scanty beard and a fatty constitution, from whose breasts so much milk flowed that he could almost have suckled an infant'); attempts at powered flight (which contains a withering put down of Leonardo da Vinci's inventive powers: 'Da Vinci ... tried it, and failed; he was an outstanding painter'); and a recipe for lip salve. *On Subtlety* meanders and stalls and jumps, but it does entertain. It's not *that* bad.

It seems that Scaliger was just looking for someone — anyone — to take down. He was, after all, a man who raised himself up on the coattails of others. Aged twelve, he had been page to the Emperor Maximilian. By the time he left the Emperor's service, aged twenty-nine, he had his eye on the position of Pope, no less. There is talk that he gave this ambition up when he could no longer bear the company of monks. He went back to soldiering (for the King of France).

Eventually, Scaliger married a girl of thirteen who bore him fifteen children. He was, according to one of his children, a 'terrible' man, more feared than loved. And his unpleasantness had not yet run its course.

Scaliger's subsequent literary career was built on attacking the public's favourite scholars. Once he had stuck his knife into the much-loved scholar Erasmus, Scaliger turned on Jerome. As the 1854 biography written by Henry Morley puts it (I won't be referring to Wykes again), 'It was a thick military book, full of hard fighting, with no quarter and no courtesy' and riddled with 'railing, jeering, and rude personal abuse'.

At first, Jerome didn't dignify the attack with a reply. In fact, his silence was so deafening that some joker told Scaliger that his vitriol had put Jerome in the grave. Taking this wag at his word, Scaliger issued an apology to the public. Scaliger's funereal oration assures Cardano's fan base (somewhat self-servingly) that 'the distress of mind occasioned to Jerome Cardano by my trifling castigations was not greater than my sorrow at his death'. This was everybody's loss, he said: 'the republic of letters is bereft now of a great and incomparable man'.

Scaliger goes on for pages and pages. And then, in what must have been a mortifying moment, Jerome published a polite and erudite reply to Scaliger's criticisms without even mentioning his assailant by name.

'That was a nice move,' I say. 'You didn't even give him the satisfaction of a citation.' With a grin, I swipe my hand away towards the door. 'Scaliger, you are dismissed.'

Jerome looks puzzled again. 'It wasn't meant to be dismissive,' he says. 'Men of letters simply ought to act with dignity in their dealings with one another.'

Jerome's dignity certainly did Scaliger no favours. Most contemporaries agreed Scaliger suffered a heavy defeat. An eighteenth-century literary scholar called Tiraboschi likened the dispute to a fight between a giant and a girl.

Jerome shrugs. 'I have known girls who fight well,' he says. 'My mother, for example.'

It's not that Jerome is impervious to assault. Later slurs on his character and talents hit him hard, especially when they took hold in Pavia. Though Milan had always caused Jerome grief, the people of Pavia had always been kind and respectful to Jerome — it was, to him, a place of refuge. But, after Giovanni's execution, the mud of Milan still clung fast and there were mutterings among the Pavia scholars about whether Jerome's position should be reconsidered.

The Milanese scholars had played a perfect hand. It is in 1562 that Jerome first learns that Scaliger had not been the only one who was out to get him. Some of the Milanese senators who condemned his son to death two years ago have been bragging about doing so in a deliberate attempt to drive Jerome mad with grief. Writing of this time, he claims to 'not know whether I was most wretched or most hated'.

He is not destitute yet, though. He still has a house and a job, at least: he is a Professor at the University and, as a member of the Pavian College of Physicians, he can practise medicine to recoup some of the savings that are now gone. However, Jerome's lifestyle is making his position even more precarious. His three young lodgers are inciting gossip. They are handsome boys, given to music and gambling — passions that Jerome shares and encourages. They are whispered about as an immoral household. The faculty at Pavia is starting to turn against him.

Jerome's solution is to seek employment in Bologna. There, he approaches Carlo Borromeo, a family friend who has risen through the clerical ranks to become a cardinal — indeed a saint. Borromeo agrees to help, as does Cardinal Morone, another old friend. With an appointment in Bologna all but secured, Jerome offers his resignation to the University of Pavia. The Senate, interpreting the offer as a rash act by someone in an agitated state of mind, refuses to accept it.

The intrigues are not over. One morning, a courier arrives at his Pavian house with a letter. It purports to be from his daughter, Chiara, and her husband, 'a most infamous and vile letter,' Jerome calls it in his autobiography. In it, the couple claim to be ashamed of their relationship with Jerome and suggest that Pavia's Senate and the College should sever all relations with him in reaction to the appalling acts he has committed. Then another condemning letter comes. This is from a colleague at Pavia named Fioravanti. As Jerome recalls, the tenor of it is that 'he was ashamed of me for the sake of his country, the college, and the faculty; that the rumour was being circulated everywhere that I was using my boys for immoral purposes; and that not satisfied with one, I had added another to my household — a state of affairs absolutely unprecedented'.

Jerome dons his cloak and sweeps out of the house to confront Fioravanti. The confrontation works: Fioravanti breaks down and confesses that he wrote the letter purporting to be from Jerome's daughter and son-in-law, and that he did so on behalf of the rector at the University of Pavia. He also says an accusation is coming, nonetheless. Immediately, Jerome works out what is going on. It is another academic tussle. A man called Delfino, a friend of the rector of Pavia University, has designs on Jerome's position.

The accusation, when it comes, is — predictably — of impropriety with the boys under his roof. Jerome is pulled into slanging matches in the street, writing later of the ring of onlookers attracted by the war of words. The repercussions reach Bologna, where Jerome's enemies side track an emissary from the university there and tell him that the old man is unworthy of a job. He is not even popular as a teacher, they say — he lectures to empty halls in Pavia. And so reports reach the Bolognese Senate that Jerome is 'a professor without a class, but only benches; that he is a man of ill manners, and disliked by all; one full of folly. His behaviour is repulsive; and he knows but little of the art of medicine, expressing such sectarian opinions about it that he is rejected by all in his own city, and has no patients.'

On hearing the report, the Senate at Bologna halt the recruitment process. Fortunately for Jerome, Cardinal Borromeo knows the report is false, at least in part. Jerome, Borromeo tells the Senate, once cured his mother when all other physicians failed. Another member of the Senate confessed he had once been treated — and treated well — by Jerome and that he knows of other respected men who can say the same. The Senate quickly realises the reports are part of a conspiracy. They can't help wondering if there is something behind it, though, and exercise caution: Jerome is offered a short-term position. He can work for the university for a year, during which he must prove himself. Only then can he begin negotiations over a better salary.

Humiliated by the offer, Jerome refuses the terms. But not without regret. His tenure at Pavia has ended; in the wake of the accusations, that refused resignation letter was suddenly accepted. He is running out of money and of ways to earn it. He becomes ever more reclusive and paranoid — not a bad thing, as it turns out. Thinking his enemies' next move might be to present his writings in a dark, twisted light to the growing spectre of the Inquisition, he pre-empts them by sending all his published works to the Council at Rome. It is a move that might well have saved his life.

However, that good news is still some way down the line. Here in Pavia, things are about to get worse. Having resigned from the University of Pavia and turned down the offer from the University of Bologna, Jerome is desperate and destitute. When four Milanese senators write to him suggesting he is a good candidate for a professorship there now, he has no choice but to start the application procedure. And then he is forced to halt it. The Milan Senate's representatives report that two of their physicians have told the university that they have witnessed Jerome perpetrating 'grave crimes'. Out of respect, he will not be arrested, the Senate tells him. That is, not unless he strays within Milanese territory.

For three weeks, Jerome is an exile everywhere. A shadow hangs over him in Pavia, Milan, and Bologna. But he still has powerful friends, a number of whom work hard to refute the charges and ridicule his accusers. Suddenly, for reasons that Jerome will never discover, the charges are dropped. Bizarrely, the process has gained him a certain notoriety and Jerome finds himself a celebrity. 'I grew in fame,' Jerome later recalls. 'The citizens, indeed almost the whole state, embraced me with peculiar love, admired my innocence, and pitied my misfortunes.' Cardinals and councillors sought him out: 'I never met with a success greater or more splendid.'

He even receives a new, unrestricted offer from Bologna. He accepts, deliberately turning his back on the Milanese. 'I knew of nothing worse than to endure life surrounded by the cruel faces and hard voices of the men who had torn from me my sweetest son,' he says.

Even the spirits are kind: he has a dream in which he is told to put an emerald in his mouth whenever he wants to forget the pain of Giovanni's execution. Emeralds, he knows, are a means to call information from the future; perhaps his mind will be so busied with prophecy that the troubles of the past will fade from memory. From that morning until the end of his life, he is barely without that green gem tucked under his tongue. It is removed only for eating and talking.

$$\psi$$

Jerome will live in Bologna from 1562 until his arrest in 1570. This is where he takes on a student: Rudolf Silvestri, the skinny young man who will care for him in prison. He is also overjoyed to find that his former pupil, Lodovico Ferrari, is one of the university's lecturers, teaching mathematics.

Within a year, Ferrari will be dead at his sister's hand, but Jerome doesn't know this, of course (where are the signs and omens when they could be useful?). And it is good that he doesn't: these are happy times. The only taint is from the continued jealousy of academic colleagues furious at the ease with which he walked into the job.

In petty spite, the University of Bologna administrators timetable Jerome's lectures for after the dinner hour — the graveyard slot, when students are unlikely to attend. The lecture hall is frequently (and deliberately) double booked so that he is always competing with other teachers for the space. Secret letters are sent to Jerome's powerful friend Cardinal Morone, repeating the allegation that Jerome lectures to almost empty benches. For a while, Jerome also becomes embroiled in a very public row with a particularly prickly colleague: Fracantiano, the Professor of the Practice of Medicine. It is the age-old tale: Jerome is Professor of the Theory of Medicine, and a clash between Theory and Practice is almost inevitable. Indeed, Fracantiano so hates his rival that he cannot bear to be in Jerome's presence, instructing his attendants to warn him if Jerome should ever come near. Students being students, it becomes their favourite sport to bring Jerome to wherever Fracantiano is teaching. On one occasion, they succeed in enticing Jerome into one of Fracantiano's lectures and are rewarded with a wonderful sight. The Practice Professor is in such a rush to evade the Theory Professor's presence that he trips over his long black gown and falls facedown on the floor.

Years pass without anything more significant happening.

And then comes the arrest.

ψ

Jerome is in a conciliatory mood tonight.

'It is not hard to imagine how your man Wykes could have seen Aldo betraying me to my Inquisitors,' he says. 'Aldo has been nothing but trouble. It is a matter of public record that I had *him* arrested a few months ago. And that wasn't the first time.'

Jerome had been paying off his son's gambling debts, but that didn't solve the problem. Tired of having his own money and possessions disappear, he threw Aldo out of the Bologna house. Even that wasn't enough. Conspiring with one of his father's assistants, Aldo performed an audacious burglary, sneaking into the house, smashing open an ironbound strongbox and carrying away cash, jewellery, amulets, and precious stones. Jerome reported the crime, and his suspicions about the perpetrators, to the authorities.

'My assistant went to the gallows. Eight spells in a prison cell had done nothing for Aldo, so I had him banished.' There is a wry smile on his face. 'There are some perks to being a freeman of the city.'

'And now the freeman is in prison.'

He sighs, a resigned look spreading across his features. 'I imagine there are many people enjoying that irony.'

Chapter 14

The bigger they are, the harder they fall. That is one way to interpret the fact that large objects, such as you and I, don't appear in two places at once. To put it another way, maybe we find the quantum world so weird because, in our world, everything is so big that it has significant consequences and repercussions that stop the 'weirdness' in its tracks.

That's certainly what Roger Penrose thinks. Penrose, a mathematician at the University of Oxford, doesn't dispute that superpositions collapse somehow. But although he believes there is a real world, physical process behind the collapse, he doesn't think it is to do with leaking information.

For Penrose, it all comes down to gravity.

Gravity is a mysterious force. It creates a weak pull between any two objects. We know the pull is proportional to their masses, but no one knows how that actually works. All we know is that it exists and that a simple formula describes just how strong the attraction is. And we've known all that for a long time — Newton worked it out.

At the beginning of the twentieth century, Einstein reformulated Newton's gravity. He showed mathematically that the pull can be interpreted as a result of geometry. Something that has mass distorts the space around it (it also distorts time, but we'll get to that) in a 'gravitational field'. The classic illustration is the effect of a heavy object sitting on a rubber sheet. The rubber gets distorted, and anything else moving on the sheet will move not in a straight line, but along a curve determined by that distortion. Similarly, a planet moving through space distorted by the sun's mass will move in a curve — an orbit.

Einstein's theory is called general relativity and it works very well, fitting all the facts. It even explained a mystery or two, such as a strange anomaly in the orbit of Mercury that had puzzled astronomers for two hundred years. Einstein also showed that energy does the same thing as mass — they are actually equivalent. So a photon's energy will distort the space around it, just as an electron's mass will do.

Now let's go back to the double slit. We send our photon in and its wavelike properties cause it to take on a superposition. It is in two places at once. So, Penrose asks, what does its gravitational field look like?

Well, the superposition means it must now be a distortion of space that has two focuses, not one. The amount of distortion depends on the mass — if the object is light enough, or has very little energy, the distortion will be very small. Imagine holding a tapestry set loosely in a frame. You have it horizontally between your knees and place a small pebble on it. There is a tiny indentation — a distortion — where the pebble sits. Now place another small pebble somewhere else on the fabric. There will be another tiny indentation in the fabric. Make those pebbles bigger, and the indentations will be bigger. Once they get to a certain size, the indentations will merge and the pebbles will smash together. Two become one.

From here, it's not hard to see what Penrose is saying: above a certain threshold, superpositions of mass or energy will create gravitational distortions of space that can't be sustained. That, he claims, is how wave function collapse happens.

It's a neat explanation. It tells you why you can't put a cat into a superposition, for instance: it's too heavy. The cat's mass creates an extended gravitational distortion in space that means the two positions of the cat roll together into one immediately. The fabric of space simply can't keep them apart.

That's conjecture, of course; we can't do experiments with a cat. However, our best experiments have shown us that large collections of atoms seem to be unable to exist in superposition. We have made interference patterns with molecules composed of hundreds of atoms, but the more massive they get, the shorter-lived the superposition. That said, at this stage, we are talking about astonishingly massive things. Markus Arndt at the University of Vienna has created superpositions of molecules composed of more than eight hundred atoms. He and his colleagues are now working on firing viruses at a double slit and watching them produce an interference pattern on the other side, just as if they were waves, and not biological entities with a well-defined location. According to Arndt, viruses and vitamins will form superposition states. So, we haven't yet seen a quantum cat, but within a decade the kinds of big biological molecules that make up cats may well exist in two places at once.

Arndt thinks it's just a matter of engineering solutions — for him, there's no indication that a high mass directly impedes superpositions. And it is true to say that nothing in the gravitational explanation tells us why observing which slit the photon goes through would cause its superposition to collapse.

This interpretation does have one huge advantage, which is that it can be tested. And we can do more than just send ever more massive objects through the double slits. We can test the effects of the flow of time, for example.

Einstein's general theory of relativity says that clocks will run faster if they are put in a weaker gravitational field — further from the centre of the Earth, say. The notion has been tested using hyper-accurate atomic clocks. First you synchronise them, and then put them at different heights above the Earth's surface. Wait a while, then bring them back together for a comparison — and they show different times. That's why, over your lifetime, your head ages three hundred nanoseconds more than your feet. You can harness this effect to mess with an atom or photon in a double slit experiment — if the slits are at different altitudes, there should be a difference in the two branches of the superposition that affects the result you get when they recombine at the detector.

Einstein's other theory might help, too. Ten years before he came up with general relativity, Einstein created the 'special' theory of relativity. This says that anything moving through space will have a unique experience of the flow of time. It's a phenomenon that is called 'time dilation' and we know it is true because we have also tested this idea using atomic clocks. Fly two synchronised clocks around the world along two different routes, and they are out of sync when you bring them back together.

Now think of that photon moving towards the double slit. It is in a superposition, which means its time dilation is occurring along two different paths at once — it will be different on each path. Then it hits the detector and the two become one. So how much time has passed for the photon? There is a fundamental mismatch: it would now have two times (in its internal phase clock, controlled by Jerome's i, if you're wondering). That simply can't be, surely? We'll see. If we can learn to control the effect, it might have an interesting and informative effect on the interference pattern at the detector.

There is a fundamental problem with this interpretation. We know that quantum physics and Einstein's general theory of relativity are incompatible. We suspect that's because relativity does not give the final answer to the fundamental question of the nature of space and time. It works superbly as a way of probing what happens when you send a light beam or a particle through space inhabited by other massive or energetic objects. But it's a hand-waving, geometry-based illustration of some of the properties of space and time. Not how it actually *is*. For that we need a quantum theory of gravity.

Einstein spent the last decades of his life trying to unite quantum physics with relativity, a fruitless quest that saw him move in ever decreasing circles and rendered the old man an embarrassment to his colleagues. The same quest also reduced Schrödinger to a similar end. In the eyes of their younger peers, both were reduced to near ignominy.

ψ

'The world is cruel to the old,' Jerome says. There is a weariness about him tonight, as if my tales of the failing old men of physics have pulled the last vestiges of resolve from his scrawny frame. He waves his hand at the cell walls. 'You don't see young men thrown into prison for their books.'

'Young men don't often publish books,' I say. 'They keep their thoughts in their heads, where they can grow and change. It's good insurance against being defined by foolish, half-formed ideas.'

Jerome eyes me for a moment. 'How old are you?' he asks.

'Forty-six.' I know what is coming next.

'And in how many books have you set down your foolish, half-formed ideas?'

I hesitate. Five? Six? 'Four that are any good.'

He smiles. He is bouncing back. 'Anything you regret publishing?'

'One. But it doesn't matter — hardly anyone read it.'

Jerome's smile broadens, then he breaks into a gentle laugh. His shoulders shake and he wheezes as he tries to catch his breath. When he calms, he looks directly at me. Suddenly he is enjoying himself. 'Having readers is overrated,' he says. 'Look where it got me.'

Jerome never explained the reason for his arrest. That is almost certainly because he was told to keep the details of his interrogations to himself. The end result has been centuries of speculation. His biggest problem, scholars agree, was Antonio Ghislieri. In

1566, Ghislieri ascended to the papal throne as Pius V: a sombre, dogmatic, cheerless soul who had himself been a Grand Inquisitor. His zeal probably came from a sense that the Roman Catholics were on the back foot: the Protestants were ruling England and Scotland, and had strong footholds in Germany, France, and the Netherlands. Ghislieri took the step of excommunicating Elizabeth I and committed himself to stemming the tide of heresy creeping into Italy. It would, he reasoned, come from the north of the country. Jerome's life in Bologna may simply have put him in the wrong place at the wrong time.

He was also in the wrong profession. One of Pius V's edicts was that doctors should not treat patients who had let more than three days pass without confession. This stricture applied even to bedridden patients who presumably had little opportunity — or appetite — for heinous iniquities.

Jerome had ignored this law, as any number of sly, ungrateful patients might have testified. Then there was the Pope's decree that the Inquisitors should rake over the last two decades of writings by leading figures. That Jerome had already submitted his books for scrutiny no doubt helped. However, that was under a less stringent regime. Under Pius V, Jerome's stances on various doctrines were bound to suggest heresy.

There was, for instance, the passage in *On Subtlety* that compared and contrasted the views of Jews, Muslims, and Christians. There, Jerome's take was that God would be rather lenient on anyone who was trying to exercise faith of any sort and unlikely to condemn someone to the fires of Hell just because they followed a different set of doctrines. He even praises the opposition.

'The Mohammedans themselves have strong points,' he says, for 'they refrain from slaughter, dice, adultery, and unprincipled deeds against God, and unspeakable blasphemies, four faults by which the whole Christian population is almost crushed.' He lauds them for their focus on a single deity, uncomplicated by the idea of Father, Son, and Holy Spirit, and for their straightforward, honest reputation. 'What if you look at the good reputation of their women, and their mosque worship?' he asks.

It is likely Jerome's religious tolerance is born from his admiration for the mathematical insights of men such as Mohammed ibn al-Khwārizmī, the author of the ninth-century text *Al-jabr wa'l muqābalah* (from which we have derived the word algebra). Jerome revered al-Khwārizmī enough to mention him in the very first chapter of his own contribution to algebra, *The Great Art*. The thinking seems to be straightforward: if the pursuit of mathematical and scientific knowledge was nothing less than the pursuit of the mind of God, how can a devout Mohammedan mathematician be any lower in God's eyes than a Christian one?

Then there was Jerome's discussion of the immortality of the soul. This is dangerous ground in a time when a fear of eternal punishment was a useful means for suppressing the growing dissent against the church. Jerome's argument in support of the doctrine is pragmatic in the extreme: people, he says, are incapable of being good without the spectre of eternal punishment hanging over them. Therefore, it is important that they be led to believe in an afterlife. It's hardly a ringing endorsement of the Holy Church's beliefs.

Or perhaps it was his astrology. Jerome's horoscopes, especially that horoscope of Christ, are a potent source of controversy. Pius V is set full against astrology, and that goes double when it is applied to our Lord and Saviour.

ψ

Jerome remains bemused by the idea that a horoscope of Jesus is any kind of heresy. 'I published it within a work on Ptolemy — and alongside Ptolemy's own horoscope of our Saviour,' he says. 'Plenty of others have done the same and always shown that our Lord was all that he claimed. What heresy is there in backing up the truth of the Gospels?'

'Ideas evolve,' I say. 'These days, horoscopes are seen as a suggestion that the stars influence their Creator. And, by casting a horoscope, you are aiming to influence Him, too.' I hesitate. 'You know John Dee?'

I know he knows Dee, a mathematician and astrologer who advised Queen Elizabeth I of England. They met almost twenty years ago, at Southwark dock on the Thames in London. Dee's notebooks record that they went together to the French ambassador's house to examine a gemstone that could draw down the power of the planets.

Jerome's brow has furrowed. 'We carried out some investigations together,' he says cautiously.

'Did you know he was arrested for casting horoscopes of Queen Mary and the Princess Elizabeth? It was considered meddling, an attempt to influence them through occult means. His colleague burned at the stake for it.'

His eyes have widened. 'And Doctor Dee?'

'Talked his way out of it.'

Jerome's features relax into a smile. 'Yes. That sounds like him.' He frowns again. 'You know he's not a real doctor?'

It is my turn to raise an eyebrow. 'The point, Jerome, is that casting horoscopes is dangerous. It looks like magic. It looks like casting spells, or suggesting mystical influences can intervene. Your horoscope of Christ is, to them —' I point to the door, 'an attempt to twist God's arm. It is a suggestion that the planets — the creation — can control the Creator.'

Jerome's expression is blank. He is staring at the door, following my still-extended finger. Eventually, the spell breaks and he raises an eyebrow at me. 'So what should I do, my guardian angel?'

His ever-constant conviction is touching. 'You still have friends in high places,' I suggest. 'Cardinal Morone, for one. And Archbishop Hamilton. Didn't he write you a letter offering help if you ever needed it?'

He smiles, opens his arms, and does a passable Scottish accent. *'If I can be of use to you in anything, with aid, service, or money, you will send word to me … the moment I have tidings of it, consider the thing done.'*

Jerome's memory astonishes me. Without further protestation, he sits back down at his desk and begins to write.

ψ

Consider the thing done. On 8 March 1571, five months after his arrest, Jerome is told to gather his possessions and get ready to leave his cell. He is to live the remainder of his days under house arrest. I don't know if it was Hamilton. He was still alive — just — and capable of colluding in an assassination plot at fifty-nine years old, so he could certainly answer a letter. But it seems unlikely, given the time that it would take to send a letter to Scotland and then another back to Rome. It's only been a few weeks since our last conversation. So maybe it was Cardinal Morone, or Cardinal Borromeo? Maybe I cannot claim to be the one who saved him by pointing him to Hamilton. Whatever the truth, this cell is to be Jerome's home no longer.

There are conditions to Jerome's freedom. Many, many conditions. First, he must abjure — that is, renounce and repudiate his heresy — in front of the assembly of Inquisitors. This is something of a let off, though. Many of those on trial have to abjure in the city square for all to see. Someone is clearly looking out for Jerome and recommending his punishment not be too severe.

Second, he is stripped of the right to teach in any of the Papal States. His days as a Professor are over.

Third, he must give up the right to publish any of his writings. No one can stop him scrawling down his thoughts, but they are not to be broadcast, lest he lead the public astray again.

Fourth, he is to deposit a sum of eighteen hundred crowns with the church and live under a self-imposed house arrest. He will receive from it an annual income of 180 crowns. This will be, to a man used to earning four times that from his professorial salary alone, a life of penury.

Fifth, he is not to speak of the Inquisition's charge or methods.

'Not even to you,' he says, eyeing me from the open door of the cell. I am sat on the straw mattress, wondering if I will see him again.

I won't.

ψ

Months later, Jerome moves from Bologna to Rome. From the little information available, it seems that this was at Cardinal Morone's suggestion. The Roman Inquisition, ever zealous under Pius V, has taken over proceedings at Bologna; Jerome will be safer in Rome, away from those who know him and might accuse him anew. He will also be better placed to negotiate with the authorities over his status. Once settled, he works hard to have the condemnation of his writings lifted and to achieve a licence to teach and publish once more. Jerome being Jerome, settling for obscurity and poverty is not an option.

It was not a good time. Morley records that the French intellectual Jacques-Auguste De Thou happened upon Jerome in Rome, 'walking about the streets, not dressed like any other person'. Some kind of Scottish kilt, perhaps? De Thou visited Jerome in his house and recalls that he was 'a madman of impious audacity, who had attempted to subject to the stars the Lord of the stars, and cast our Saviour's horoscope'.

Mud sticks, and de Thou was an enemy of the Catholic Church, and a fan and friend of Scaliger. Enough said. These are men who thought so little of Jerome that they created and broadcast the myth that he starved himself so that he would die on the date his horoscope had predicted.

Fools.

There's clearly something between Jerome and the French of that time. A French lawyer, Francois D'Ambrose, also visited, and also paints a picture of a madman. Jerome, D'Ambrose reports, was living in a room with banners on the walls that declared TEMPVS MEA POSSESSIO — 'Time is my Possession'.

But things do improve. On 1 May 1572, just over a year after Jerome's release, Pius V succumbs to a rampaging cancer. Within twenty-four hours, the conclave has chosen Cardinal Ugo Boncompagni as his successor. Boncompagni takes the name Gregory XIII and ushers in a host of reforms, including a revision to the list of forbidden books. Gregory XIII is a lover of astronomy — a passion that led him to instigate the reform of the calendar, despite widespread opposition. What's more, Gregory has an illegitimate son, which surely makes him a little more sympathetic to Jerome's plight. Even better, the new Pope counts Cardinal Borromeo among his former students, and Cardinal Morone among his friends.

The stars have aligned. By the end of October 1572, the seventy-one-year-old Jerome is in possession of a licence to republish his existing medical works. A further eighteen months of petitioning grants him the right to publish something new. Then, in September 1574, he is accepted into the Roman College of Physicians. The Pope grants that the interest on his eighteen hundred crown bond be paid to him as a papal pension. It works out to be slightly less than his 180 crowns per month, but it is a valuable public acknowledgement of his rehabilitation, which is worth money in itself. At the beginning of 1576, Jerome wins the right to return to his teaching post in Bologna.

He never makes the journey.

ψ

There is hardly any more to say. What little I can find about the end of Jerome's life is scraped together from a variety of meagre sources. There is little of these years in his autobiography, which was published after his death. In my more whimsical moments, however, I can at least read myself into those pages. 'It has been my lot to be attended by a good and compassionate angel,' Jerome says. Remember how I wondered whether Rudolf Silvestri ever saw me? Clearly, he did, because Jerome admits it. 'In prison,' he writes, 'my guardian appeared to me and to my youthful associate … in order that he might confirm me in my hope of divine favour whereby I should escape death, and that all those trials which I was suffering should seem less hard to bear.' In his autobiography's last chapter, which revels in the utilitarian title 'And This Is The Epilogue', Jerome describes himself as 'a teller of the truth, an upright man, and indebted for my powers to a divine spirit'.

The more sceptical among you will point out that I read those words before I conceived the conceit in these pages. But because you are humans and in possession of a brain that has not evolved beyond entertaining superstition and mystery, some of you will be more open to the idea that I really did visit Jerome, that we really are somehow connected. Vishal, my astrologer, would have no problem with the concept, I suspect. After all, there is even evidence of our conversations about the nature of the universe. Jerome ends the book describing his source on such matters: 'for all things more than mortal there is my attendant spirit which can neither be described nor alluded to and is not under my control'. He wonders on the page whether what he calls his 'boundless love of truth and wisdom' drew this spirit to him. His next suspicion may be more accurate: 'perhaps my angel was present with me for an end known to himself alone'.

Perhaps. In the spirit of quantum theory, I'll leave the interpretation of these matters up to you, the observer.

And This Is The Epilogue

'An evil fate wills it that men will from time to time revert to darkness out of boredom with light. Ours is such a time, with great opportunities to learn the right things being spurned, and a wealth of the most lucid truths being disregarded in favour of obscure trivialities.'

That was how Gottfried Leibniz, mathematician and philosopher, expressed his frustrations at the beginning of the eighteenth century. You might think he was complaining about people rejecting the rational lines of enquiry that were born in the Enlightenment — people persisting with astrology, perhaps, in an age when astronomy was in the ascendant. In fact, the jibe was aimed squarely at Newton, and his weird and occult idea of 'forces'. The complaint comes in a treatise Leibniz called 'Against Barbaric Physics' directed at those who 'speak of attracting and repelling, adjusting, expanding and contracting forces'. Leibniz admits that the idea of the planets 'gravitating and striving toward each other' has been confirmed as correct. But he rejects the new idea that 'matter is supposedly able to perceive and covet even things which are remote'. Physics stripped of all mysticism is 'too clear and simple for these people,' Leibniz complained. 'Instead they revert to fanciful ideas.'

Leibniz compares Newton to those who suggest a foetus is formed by intelligent spirits that attend to it in the womb. 'How can any reasonable person today subscribe to a belief in fantastic qualities that is tantamount to a betrayal of all natural principles? ... all this is bound to lead us completely into the realm of the obscure ...' Everything, Leibniz says, should ultimately be reduced to mechanisms, for mechanics.

Leibniz would be horrified at where we have got to today. David Bohm suggested that quantum mechanics is a misnomer: it should be called quantum non-mechanics because everything we have learned says there is no physical mechanism to be found within the theory. There is certainly no mechanism that explains entanglement. There is no physical conception of what happens when the photon approaches the double slit. And as for Newton's gravity, a proper mechanical explanation still eludes us.

Einstein reworked Newton's gravity as a set of contours in space and time: we can predict the trajectories of the planets if we consider them as traversing a landscape of hills and valleys created by the mass and energy contained within the universe. But we sometimes confuse our success in explaining what happens with an explanation for why or how it happens. Einstein's work has proved an extraordinary success, with myriad successful predictions and explanations of physical phenomena under its belt. But it is not an explanation of why and how mass and energy should give shape to the cosmic landscape — or even that the shape is really there in the way we say it is. After all, a deformation in a two dimensional rubber sheet requires a third dimension. If we are distorting the four dimensions of space and time, into what dimension are they distorting? Don't misunderstand me — Einstein's explanation is wonderful. But that doesn't mean the explanation is actually how it is. Where gravity is concerned, we haven't progressed much beyond 'red sky at night, shepherd's delight'.

The truth is that quantum mechanics and relativity remain our best theories, yet also provide our most frustrating impasse. We present them as useful schema for explaining observed phenomena, but only to a point. We cannot actually account for the details of either. 'Nobody knows how it can be like that,' Richard Feynman said of quantum theory. And relativity, most agree, is a theory already in need of an overhaul, for it simply doesn't give us meaningful answers to all the questions we ask of it. Hundreds of years after Jerome's investigations of the universe, the job is still only half done.

If that. Roger Penrose's 'gravitational collapse' interpretation of quantum theory is an attempt to sketch out what a theory that unites relativity with quantum mechanics might achieve, but it is barely even a sketch. It is impossible to know how far along the path to such a 'final theory' we have travelled. In Jerome's time, astronomers had made major steps forward since the times of the ancients, but to us their tiny, Earth-centred universe still looks like a primitive and wholly inadequate attempt at understanding the nature of the cosmos. Unless we live in a special time in the history of humanity — and it is unscientific to think that we do — there is no reason to believe that the humans of the twenty-fifth century will look at us any differently from how we look at our scientific predecessors. But what will they make of our inability to parse the meaning of quantum theory? Will our take on entanglement and superposition look to them as Jerome's astrological inference looks to us?

If they were to visit us in the prison cells of our ignorance, what secrets might they convey? Part of me hopes they would tell us that the astrologers were onto something. Something deep within me, deep within all humans that have ever lived, yearns to be connected to the cosmos, to be a cog, however tiny and infinitesimally unimportant, that turns because of a distant rotation elsewhere in time and space. I suppose that is part of the appeal of Gerard 't Hooft's superdeterminism interpretation of quantum mechanics: it feeds the desire to be part of something bigger. It absolves us of the duty to explain, and allows us to just be, to let Fate direct our path through the cosmos.

Not that most physicists would subscribe to superdeterminism. But neither are they comfortable with what is currently considered the best hope for understanding how the universe works: string theory.

Remember how Einstein and Schrödinger were reduced to dust by the attempt to unite quantum theory with relativity? There were times when they refused to talk to one another, so intense was the competition between them to crack this puzzle. They even spoke of lawsuits to prevent the other from stealing ideas. String theory, our current best candidate for a quantum theory of gravity is very different from anything that pair suggested, but it is proving equally divisive.

To its critics, it is so far off the mark that it is 'not even wrong' and a 'new version of medieval theology'. Those who think it worth pursuing, on the other hand, are often puzzled by others' inability to appreciate its 'beauty'.

What is string theory? A true understanding of this attempt to unite quantum theory and Einstein's relativity is only possible for those who deal in mathematics more easily than in language. So I can only tell you an approximation of its claims: string theory says that all the familiar particles arise from the different vibrational modes of strings and loops of pure energy. The energy vibrates one way and it creates a photon. Another kind of vibration gives us an electron. And so on.

The idea has arisen from work carried out in the 1920s, when even quantum theory was young. Two mathematical physicists, Theodor Kaluza and Oskar Klein, independently worked out that Einstein's relativity can give birth to the electromagnetic theory governing the behaviour of photons and electrons if the universe is allowed to have four spatial dimensions instead of the usual three. Kaluza and Klein's extra dimension had to be curled up into a tiny circle for the maths to work, but it was an interesting result that mathematicians played with for decades, slowly expanding its sphere of influence. By the 1980s, the idea had become a theory that accounted for the existence of all the fundamental particles. There was only one problem. The mathematics only worked if the string theorists invoke even more spatial dimensions. According to string theory, there are ten dimensions of space, with seven of them lying, curled up, beyond our perception. What's more the theory doesn't describe our universe specifically. It describes a plethora of universes, each with slightly different physical properties. We inhabit — hopefully — one of 10^{500} universes described by string theory.

It is a situation that some have described as nonsensical. It is, after all, a theory that doesn't make any testable predictions and so can never be tested. 'I have been brought up to believe that systems of belief which cannot be falsified are not in the realm of science,' is how Nobel laureate Sheldon Glashow has dismissed string theory. Hence his contention that it is no different to medieval theology.

Glashow probably doesn't appreciate that it is closer to medieval — or at least Renaissance — science. In some sense, Jerome's *aevum* is still with us. String theory even has a contemporary version of Jerome's centre that 'corresponds to every point on the circumference'. It is known as the 'holographic principle', and was conceived by an Argentinian string theorist called Juan Maldacena. His idea is that all of what we term physical reality results from information held on the edge of our cosmos — a cosmos that has many more dimensions of space than the three we experience. Our physical universe, in other words, is just a small manifestation of something that exists in the great beyond, something that is unreachable, and beyond our understanding.

Though many of today's physics luminaries, Stephen Hawking among them, enthusiastically subscribe to the holographic principle, there is no experimental test that can tell us whether it is right. It is simply our best stab at a final theory. We are still very much on the road towards understanding, still travelling in the dark, with still only the faintest hope of arriving at our destination. And we are almost certainly only a few steps ahead of the man who showed us the path that we should follow, my friend Jerome.

Author's Note

In case you were in any doubt, this is not an academic work. It is not something to be referenced, or consulted as if it were scholarly research. That is why I would like to introduce you to my sources, many of which do fit that description.

First, nothing compares to reading Jerome's own descriptions of his life and work, his hopes and regrets. *The Book of My Life*, translated by Jean Stoner, is wonderfully entertaining. There are some moments of self indulgence in there, but it's an autobiography — what can you expect? Also wonderful, though a lot, lot longer, is *The De subtilitate of Girolamo Cardano*, a translation from Jerome's Latin by John M Forrester edited by the University of Edinburgh scholar John Henry. There are so many gems in these pages that I feel I will be mining them for the rest of my life. I do miss Jerome now that this book is finished, but I still get to hear his voice through these two works.

The Quantum Astrologer's Handbook would never have been written without Henry Morley's exhaustive *Jerome Cardan: the life of Girolamo Cardano, of Milan, physician*. That is partly because this biography, published in 1854, informed Alan Wykes's *Doctor Cardano, Physician Extraordinary*, the book that finally tipped me into dedicating time to getting to know the real Jerome.

Wykes says Morley's book is 'afflicted with Victorian turgidity'. I don't agree, but I can see where he is coming from. Wykes's book, published in 1969, is certainly not turgid — it is an entertaining and lucid tale, great company in front of a roaring fire, a single malt in hand. But it is also, as you'll know by now, not to be trusted. That is not a condemnation, exactly. I have friends who could be described using the exact same phrasing and I would not be without any of them.

I also used a few other books on Jerome's life and work which *can* be trusted. Let's put them into chronological order:

JEROME CARDAN: a biographical study
by William George Waters

Waters' 1898 volume is essentially a distillation of Morley's work (and also afflicted with 'Victorian turgidity', according to Wykes). There's not much more to say about it, except to be thankful, in the light of Morley's magnum opus, for its relative brevity.

CARDANO: the gambling scholar
by Øystein Ore

'An all but extinguished scientific luminary of 16th-century Italy shines again,' gushed *The New York Times* on this book's publication in 1953. 'Briskly interesting to lay readers, sometimes scarcely falling short of being racy,' said the *New York Herald Tribune*. Well, 'racy' perhaps meant something different back then, but it's certainly interesting. Ore was a Norwegian mathematician. It's not clear how he came to celebrate Jerome's life in print but his dissection of Jerome's work on probability is thorough and informative.

GIROLAMO CARDANO 1501-1576: *physician, natural philosopher, mathematician, astrologer, and interpreter of dreams*
by Markus Fierz (translated by Helga Niman)

This biography was written by a Swiss mathematical physicist working from Jerome's collected works, the *Opera Omnia*, in 1977. For a physicist, Fierz is wonderfully generous towards Cardano, giving him credit for his open-mindedness and appropriate use of astrology. I particularly like his assertion that 'Cardano did not reject the Copernican theory completely.'

THE CLOCK AND THE MIRROR:
Girolamo Cardano and Renaissance medicine
by Nancy G Siraisi

We are in 1997 now — only twenty years ago! Siraisi's book is primarily concerned with Jerome's medical practice, how it fitted with the times, and how it compares with the work of today's practitioners. There is much to enjoy here, and Siraisi is generally impressed by Jerome's insight. She highlights his understanding that men could be rendered impotent by a curse, as long as they believed in it, and that magical remedies for impotence worked simply because they 'substituted hope for the patient's despair'. Siraisi's view is that Cardano was a level-headed investigator of the physician's art. 'He repeatedly urged, in a manner highly unusual in the sixteenth century … that the first thing to ask about a supposed marvel was not what had caused it but whether it had happened.'

CARDANO'S COSMOS:
the worlds and works of a Renaissance astrologer
by Anthony Grafton

Grafton's book, published in 1999, offers deep insight into the human mind's yearning for advice, however unreliable the source. My favourite part is the dissection of economics as the modern counterpart to astrology. 'Like the economist, the astrologer generally found that the events did not match the prediction,' Grafton says, 'and like the economist, the astrologer normally received as a reward … a better job and a higher salary.'

There were sources other than books. It was a revelation and an utter joy to discover that my friend and colleague Professor Artur Ekert was also somewhat taken by Jerome. In 2008, Ekert wrote a short treatise, 'Complex and Unpredictable Cardano', laying out why his discovery of probability and complex numbers linked Jerome to quantum theory. Ekert is one of the inventors of quantum cryptography, the art of using the fundamental laws of physics to encrypt information. I feel that Jerome would, in many